The Death of Superman

Dan Jurgens
Jerry Ordway
Louise Simonson
Roger Stern
WRITERS

Jon Bogdanove
Tom Grummett
Jackson Guice
Dan Jurgens
PENCILLERS

Brett Breeding
Rick Burchett
Doug Hazlewood
Dennis Janke
Denis Rodier
INKERS

John Costanza
Albert DeGuzman
Bill Oakley
Willie Schubert
LETTERERS

Gene D'Angelo
Glenn Whitmore
COLORISTS

SOMEWHERE ELSE...
KRAANG!
KRAANG!
KRAANG!
KRAANG!
...DOOMSDAY IS COMING!

UNRELENTINGLY...
KRAANG!
KRAANG!
KRAANG!
KRIINK!
...DOOMSDAY IS COMING!

UNSTOPPABLY...
KRAANG!
KRAANG!
KRAANG!
KRAAK!
...DOOMSDAY IS COMING!

UNBELIEVABLY...
KRAANG!
KRAANG!
KRAANG!
KROOM!
...DOOMSDAY IS HERE!

SOMEWHERE ON EARTH...
SQUAK! SQUAK! SQUAK! SQUAK!
KRAANG!
KRAANG!
KROOM!
KRAKOOM!

KRAAKK!
KARAAKK!
KRAKA-
DOOM!

DOOMSDAY!
PART ONE
STORY: LOUISE SIMONSON
PENCILLER: JON BOGDANOVE
INKER: DENNIS JANKE
LETTERER: BILL OAKLEY
COLORIST: GLENN WHITMORE
ASSISTANT: JENNIFER FRANK
EDITOR: MIKE CARLIN
SUPERMAN created by JERRY SIEGEL & JOE SHUSTER
HE'S SO... AWESOME! HE SAVED ME... TWO TIMES NOW. HE'S MY FRIEND.
I WISH I COULD TELL HIM... BUT... I JUST CAN'T.
MAXALEH'S HARDWARE
OPEN PLEASE COME IN
HOURS
KEYS MA
WHILE U W
ARMOR LOCK
SO THERE'S ONLY ONE THING LEFT TA DO.
CHAN-L-OCK
FLEISCHER-PAI
INTERIOR/EXT
ENAMEL & FLA
CUSTOM BLENDED
YOU SURE THIS PAINT REALLY GLOWS IN THE DARK?
YEAH. LATEST THING. GUARANTEED.
SAY, YOU AIN'T PLANNIN' TA USE IT ON ANY SUBWAY WALLS, ARE YA, KID?
NO. 'COURSE NOT.
AT LEAST... NOT EXACTLY.
HARDWARE
I'M SCARED. BUT I GUESS IT MAKES SENSE TA BE SCARED--
FLUORESCEN YELLOW

"-- WHEN THE GUYS I'M GOIN' AFTER ARE MONSTERS!"
THIS POWER STATION IS OURS!
SO... HOW DO WE GO ABOUT DIVERTING THE POWER TO OUR WAR MACHINES?
I DO WHAT YOU SAY, CLAWSTER. FOR NOW, YOU THE BOSS.
KEEP ENGINEER HYPNOTIZED, KATHANA!
YOU HEAR CHARLIE, MAN! SO TALK! TELL US WHERE IS SWITCH SO WE CAN STEAL ELECTRICITY!
CURTIS
THE MAIN CONTROL BOARD... IS OVER THERE!
FOOD, JUICE, FLASHLIGHT, EXTRA BATTERIES, SPRAY PAINT...
CHILDRENS' A
COATES CHIL
DRENS' AID CIETY
ILDRENS CENTER
WHY IS IT I KEEP THINKIN' I'M DOIN' SOMETHING DUMB?
MAYBE I BETTER GO OVER MY PLAN ONE MORE TIME?
THE MONSTERS IN THE SEWERS SAY THEY GOT MY MOMMA...
...AN' THEY'LL KILL HER IF I TELL ANYBODY THAT THEY WANNA MAKE WAR ON THE CITY!
BUT... WHAT IF THEY'RE LYIN'? WHAT IF THEY'RE TRYIN' TA TRICK ME?
WHAT IF THEY DON'T HAVE MY MOMMA AT ALL?

BUT--THEY CAN'T BE LYIN'-- CAN THEY?
MOMMA DIDN'T COME BACK TA GET ME FROM THE ORPHANAGE 'CAUSE SHE COULDN'T.
'CAUSE SHE'S THE MONSTERS' PRISONER.
SHE WANTS TO COME, BUT SHE CAN'T. BUT I'LL SAVE HER.
AN' SHE'LL HUG ME AN' TELL ME HOW BRAVE I AM AN' ASK ME TO STAY WITH HER ALWAYS...
...AN' I'LL BRING HER OUT AN' THEN I'LL TELL SUPERMAN 'BOUT THE MONSTERS' PLAN.
THERE'RE GONNA BE A LOT MORE FORKS IN THE TUNNELS DOWN HERE. I HOPE THIS WORKS.
PSSSR
AWRIIIGHT! MOMMA AN' ME CAN'T GET LOST NOW!
HI, LOIS! HOW'RE THE ARTICLES ON THE CONFISCATED ELECTRONICS THE COPS "LOST"?
DAILY PLANET
WELL ENOUGH. MY SOURCE IN THE HARBOR PATROL TELLS ME-- WHAT'S THIS?
SEND SUPERMAN TO THE BASEMENT UNDER THE WEST SIDE POWER STATION. METROPOLIS IS IN DANGER!
-A FRIEND
CLARK ISN'T AROUND. DARN. SOMETIMES HE FORGETS TO CHECK HIS MAIL...
...ESPECIALLY WHEN HE HEADS DOWN FROM THE ROOF.
"ANONYMOUS TIP SAYS THERE'S DANGER IN THE BASEMENT OF THE WEST SIDE POWER STATION. MEET ME THERE! LOIS"
THAT SHOULD DO IT!
LOIS--?
GOTTA RUN, FRAN! WOULD YOU TELL CLARK THERE'S A MESSAGE FOR HIM ON HIS COMPUTER?

ELSEWHERE...
CHEEP CHEEP
CHEEP
CHEEP
CHEEP CHEEP
CHEEP?
BLORCH
HAH...HA HA HAAA

SCANNED THE CITY, BUT DIDN'T SEE ANY UNDER-WORLDERS ON THE SURFACE.
MUST BE LYING LOW FOR NOW. MAYBE IT'S TIME --
"--I GOT BACK TO WORK!"
HI, CLARK. LOIS LEFT YOU A COMPUTER MESSAGE.
VERY HIGH TECH OF HER. THANKS, FRAN.
TAK-TAP-TAK-TAP!
READ : MESSAGES
WHAT NOW?
ALL THE COMPUTERS DON'T WORK?
LIGHTS, EITHER!
DON'T TELL ME... WE'RE HAVING ANOTHER BLACKOUT?!
WE DO IT!
YEAAAA!
NOW CITY IS HELPLESS... AND UNDERWORLD CAN KICK BUTT!
WE STEAL METROPOLIS'S ELECTRICITY!

METROPOLIS POWER & LIGHT NEW TROY POWER STATION DIVISION OF LEXCORP
DANGER KEEP OUT!
GOOD THING SOMEONE INVENTED WIRE CUTTERS. I'D'VE HATED TO'VE HAD TO USE MY TEETH.
BLACKOUT MEANS THE ALARMS DON'T WORK. THAT'S THE GOOD NEWS.
BUT NOBODY'S AROUND, EITHER...
...WHICH MEANS THEY'VE BEEN TAKEN CARE OF.
NOTE WAS RIGHT. SOMETHING'S UP. BUT WHAT?
MAYBE DOWN-STAIRS...
WHAT'S THIS--?
THUDD!
OUCH!
IT'S LIKE KICKING CONCRETE!
HAR HAR HAR! YOU NOTICE, HUH?
OUTSIDE AND IN, THROUGH AND THROUGH, CLAWSTER INVULNERABLE.
WHO YOU BE, ANYWAY?
SHE'S THAT REPORTER DAME, LOIS LANE.
YOU KNOW HER?
SHE VOLUNTEERS AT A SOUP KITCHEN WHERE I USETA EAT.

WHO ARE YOU? WHAT'S GOING ON? CHARLIE-- HOW COULD YOU?
IT EASY, REPORTER LADY. AN' SMART. YOU HELP TEACH CHARLIE, DIDN'T YOU?
HE SOCIETY'S REJECT... LIKE US ALL.
YOU WANT TO GIVE US HANDOUTS!
BUT UNDER-WORLD GOTS WAR MACHINES NOW... AN' WE TAKE WHAT WE WANT!
SHE MIGHT BE USEFUL AS A HOSTAGE, CLAW. BETTER BRING HER TO THE BOSS.
YOU TEASE HER, CHARLIE. YOU KNOW BOSS DON'T TAKE PRISONERS!
SHE BE DEAD 'FORE HOUR IS UP!
NO PRISONERS? BUT WHAT ABOUT MY MOMMA?
THEY... THEY NEVER HAD MY MOMMA! THEY LIED TO ME SO I WOULDN'T TALK!
I-I GUESS... I KNEW... ALL ALONG THEY DIDN'T HAVE HER. I JUST WANTED TO BELIEVE...
...THAT I COULD FIND HER... AN' SAVE HER...
...AN' MAKE HER LOVE ME.

ELSEWHERE...
KRAK KRAM!
KRAKK!

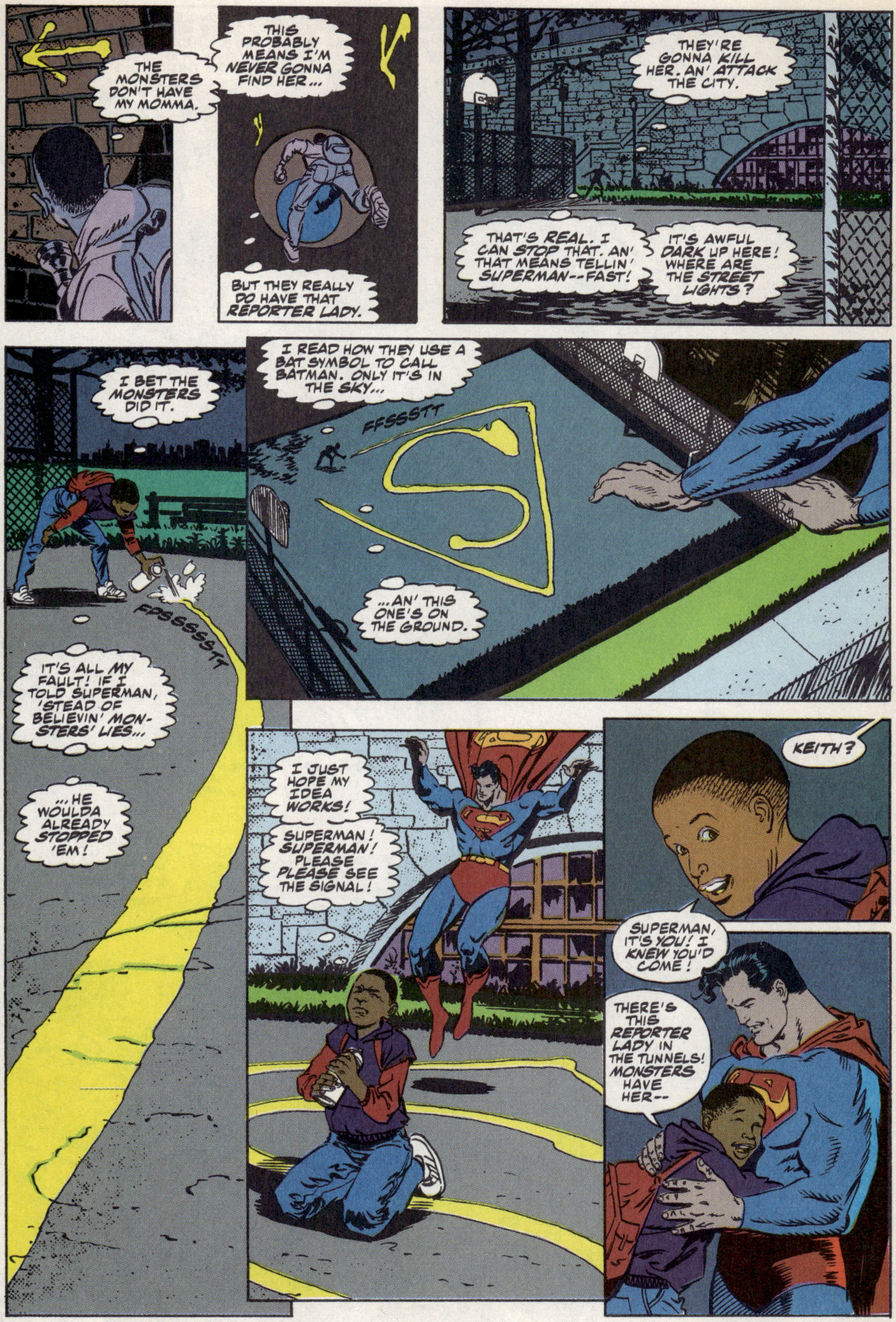
THE MONSTERS DON'T HAVE MY MOMMA.
THIS PROBABLY MEANS I'M NEVER GONNA FIND HER...
BUT THEY REALLY DO HAVE THAT REPORTER LADY.
THEY'RE GONNA KILL HER. AN' ATTACK THE CITY.
THAT'S REAL. I CAN STOP THAT. AN' THAT MEANS TELLIN' SUPERMAN--FAST!
IT'S AWFUL DARK UP HERE! WHERE ARE THE STREET LIGHTS?
I BET THE MONSTERS DID IT.
FPSSSSSSTT
IT'S ALL MY FAULT! IF I TOLD SUPERMAN, 'STEAD OF BELIEVIN' MONSTERS' LIES...
...HE WOULDA ALREADY STOPPED 'EM!
I READ HOW THEY USE A BAT SYMBOL TO CALL BATMAN. ONLY IT'S IN THE SKY...
FFSSSTT
...AN' THIS ONE'S ON THE GROUND.
I JUST HOPE MY IDEA WORKS!
SUPERMAN! SUPERMAN! PLEASE PLEASE SEE THE SIGNAL!
KEITH?
SUPERMAN, IT'S YOU! I KNEW YOU'D COME!
THERE'S THIS REPORTER LADY IN THE TUNNELS! MONSTERS HAVE HER--

"--AN' THEY'RE GONNA INVADE THE CITY!"
SOON... SOON OUR BORER WILL CARVE A SUBTERRANEAN HIGHWAY THROUGH THE BEDROCK!
THEN OUR WAR MACHINES ROLL UP THROUGH TUNNEL...
...AN' METROPOLIS WILL BE IN HANDS OF MONSTERS!

SKRRRSHRRZAM!

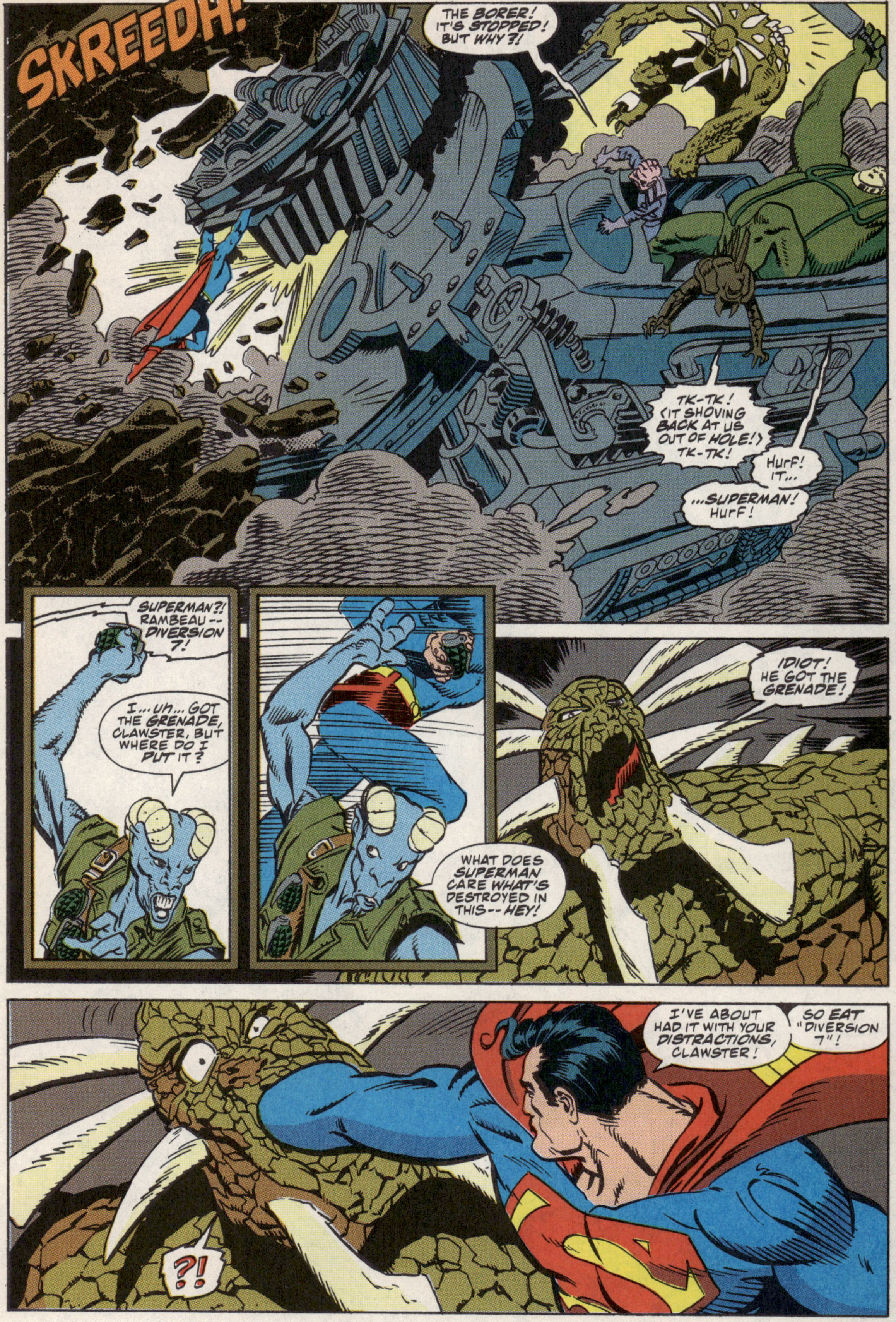
SKREEDH!
THE BORER! IT'S STOPPED! BUT WHY?!
TK-TK! (IT SHOVING BACK AT US OUT OF HOLE!) TK-TK!
HURF! IT...
...SUPERMAN! HURF!
SUPERMAN?! RAMBEAU-- DIVERSION 7!
I... UH... GOT THE GRENADE, CLAWSTER, BUT WHERE DO I PUT IT?
WHAT DOES SUPERMAN CARE WHAT'S DESTROYED IN THIS-- HEY!
IDIOT! HE GOT THE GRENADE!
I'VE ABOUT HAD IT WITH YOUR DISTRACTIONS, CLAWSTER!
SO EAT "DIVERSION 7"!
?!

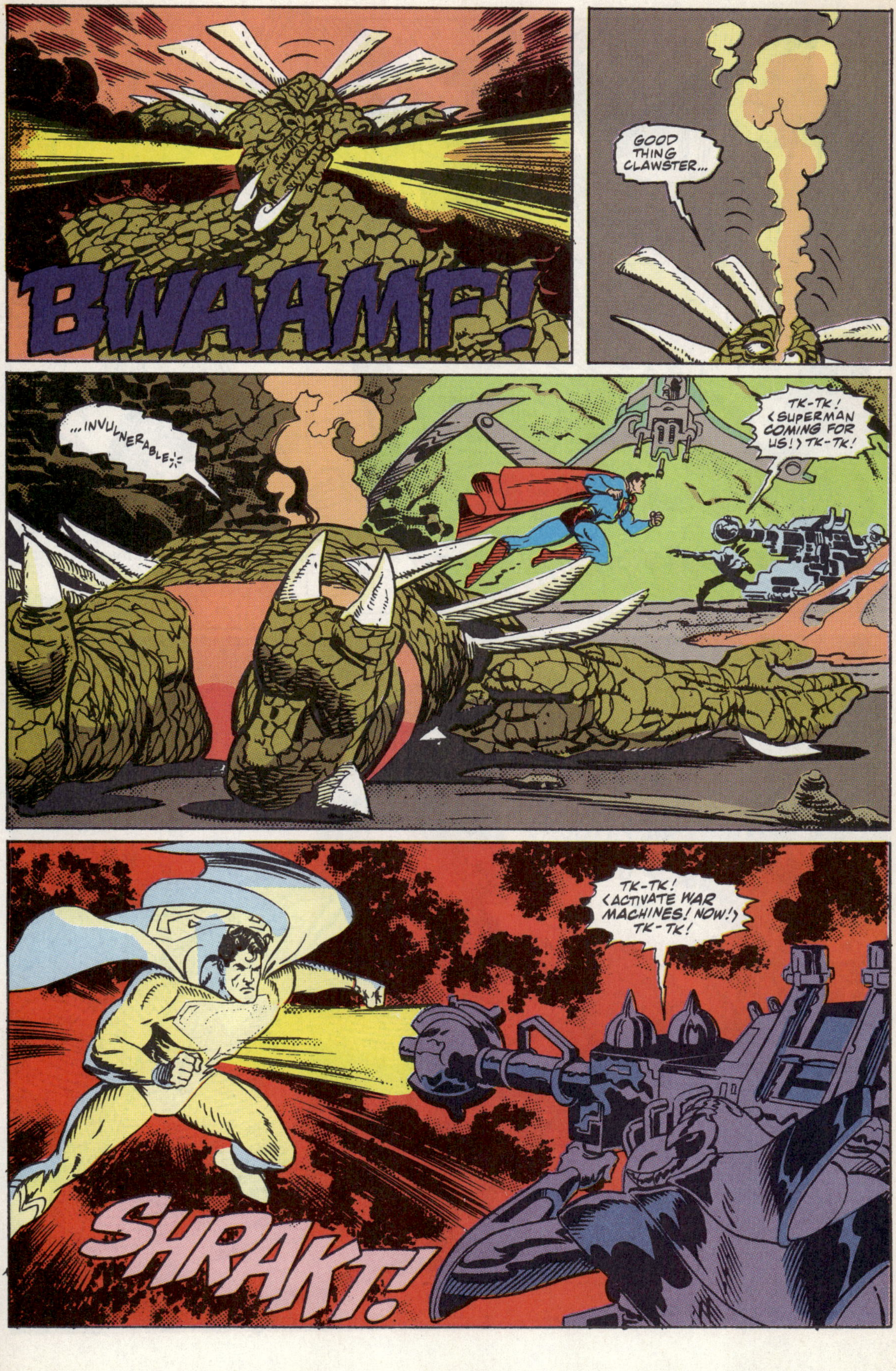
BWAAMF!
GOOD THING CLAWSTER...
...INVULNERABLE!
TK-TK! (SUPERMAN COMING FOR US!) TK-TK!
TK-TK! (ACTIVATE WAR MACHINES! NOW!) TK-TK!
SHRAKT!

TK-TK! ‹THAT THE WAY! THESE ›TK‹ WAR WORLD DESIGNS!› TK-TK!
YEAH! POUR IT ON!
EVEN SUPERCREEP CAN'T STAND AGAINST THESE WEAPONS!
AS I FEARED... THE WAR WORLD ESCAPEES ARE BEHIND THIS! GOT TO...
BRAKADRUMVF!
...GET TO MY FEET!
PINCER... COMING TOWARD ME... MEANT TO IMPALE ME...
RRRPT!
...BUT I'M NOT THE GUY WHO'S GETTING THE WRONG END OF THE STICK!

MISSED MY CHANCE TO STOP THEM ONCE...
WHAKT!
WON'T MAKE THAT MISTAKE AGAIN!
‹WHUFF!›
‹HUFF!›
‹THAT--›
‹--SMARTS!›
POUND!
KRAZZZZH!
KRL'LI'S UNCONSCIOUS! THE BATTLE SEEMS TO BE GOING AGAINST US!
WHAM
PERHAPS, AFTER ALL, DISCRETION WOULD BE THE BETTER PART OF VALOR!
COME BACK HERE, WAR WORLDER!
WHERE'S THE LADY REPORTER YOU'RE HOLDING CAPTIVE?
C-CAPTIVE? SURELY, SUPERMAN, YOU KNOW... WE WARWORLDERS DON'T TAKE CAPTIVES!
WHERE IS SHE?
CHARLIE TOOK HER TO THE EAST TUNNEL! BY NOW, SHE'S DEAD!
W-WHAT ARE YOU GOING TO DO?
POUND YOU UP TO YOUR NECK INTO THE ROCK FLOOR!
AND MAYBE, IF YOU'RE WRONG, I MIGHT JUST BOTHER TO DIG YOU OUT AGAIN!

ELSEWHERE...
HUNF?
HAR!
KRAK-KAK!
KA-DOOM!
HA-HA-HA-HA!

B-DAM
LOIS! THEY SAID YOU'D BE HERE... THAT YOU'D BE DEAD! ARE YOU ALL RIGHT?
I'M FINE!
WHAT?!
SUPERMAN! DON'T--!
DON'T YOU DARE PUNCH OUT CHARLIE! HE'S THE REASON WE KNEW ABOUT THE UNDERWORLDERS' SCHEMES!
WE DID?
SURE! YOU'RE HERE, AREN'T YOU? WELL, CHARLIE'S OUR INFORMANT!
HE'S A HERO! HE SAVED METROPOLIS!
RIGHT?
HE WAS A PACIFIST IN THE '60S AND HE WAS IN TOUCH WITH THE ANTI-WAR SEGMENT OF UNDERWORLD.
AND HE DECIDED TO STOP THEIR WAR AGAINST THE SURFACE.
I WROTE HER A NOTE TELLIN' HER TO WARN YOU, AND, WELL, I THOUGHT MAYBE IT WOULD HELP HER GET A STORY...
...PAY HER BACK FOR ALL THE DONUTS SHE'S BOUGHT ME THIS YEAR.
I KNEW SHE'D GET THE MESSAGE THROUGH TO YOU!
THE COMPUTER... RIGHT?
HAD THE HARDEST TIME CONVINCIN' 'EM TO LET ME BE THE ONE TO... SUPPOSEDLY BLOW HER AWAY.
NEVER OCCURRED TO ME SHE'D COME DOWN HERE AND GET HERSELF CAPTURED.
HOW ELSE WOULD SHE GET THE STORY?
AND NOW... IS ANYBODY GOING TO UNTIE ME?

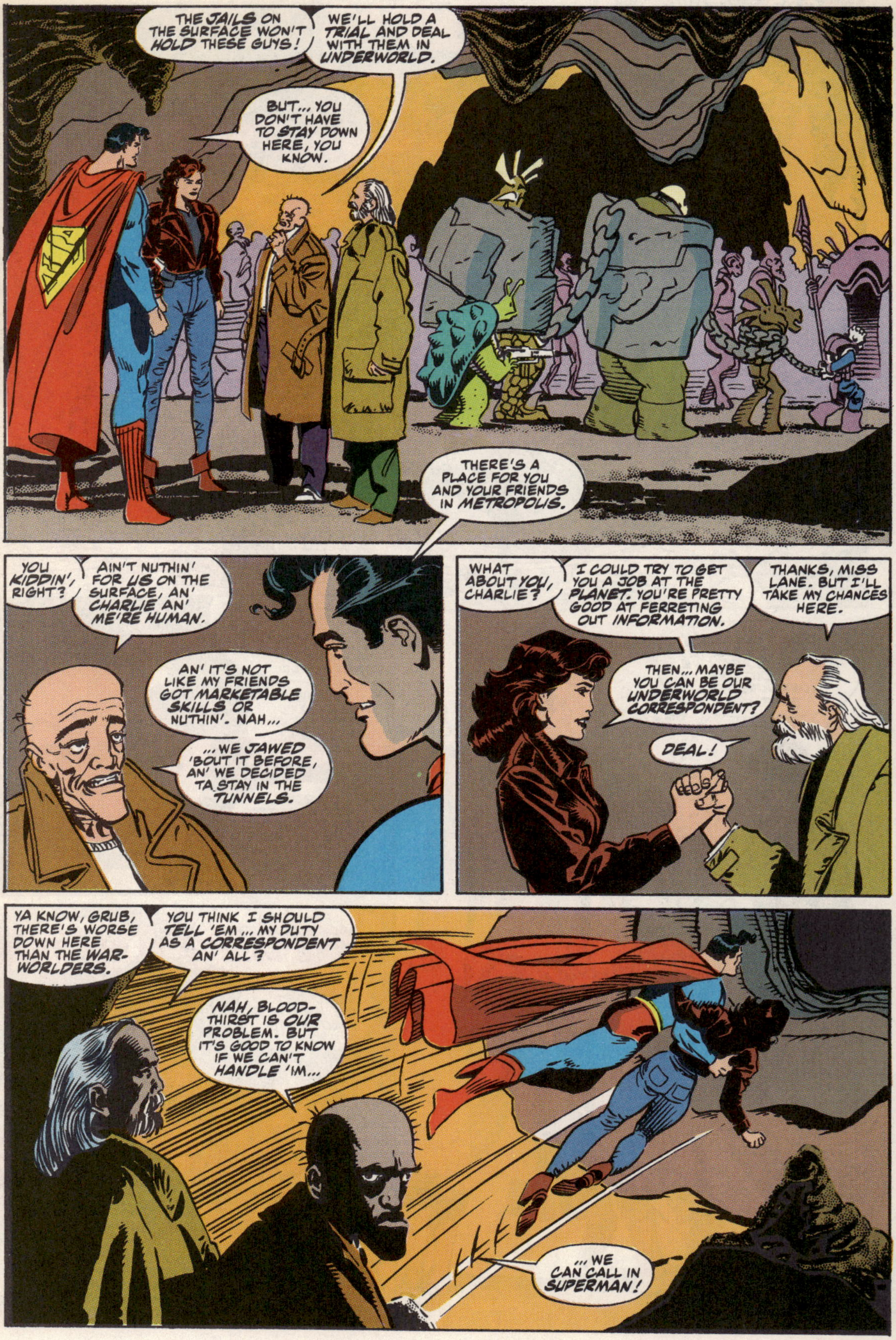
THE JAILS ON THE SURFACE WON'T HOLD THESE GUYS!
WE'LL HOLD A TRIAL AND DEAL WITH THEM IN UNDERWORLD.
BUT... YOU DON'T HAVE TO STAY DOWN HERE, YOU KNOW.
THERE'S A PLACE FOR YOU AND YOUR FRIENDS IN METROPOLIS.
YOU KIDDIN', RIGHT?
AIN'T NUTHIN' FOR US ON THE SURFACE, AN' CHARLIE AN' ME'RE HUMAN.
AN' IT'S NOT LIKE MY FRIENDS GOT MARKETABLE SKILLS OR NUTHIN'. NAH...
...WE JAWED 'BOUT IT BEFORE, AN' WE DECIDED TA STAY IN THE TUNNELS.
WHAT ABOUT YOU, CHARLIE?
I COULD TRY TO GET YOU A JOB AT THE PLANET. YOU'RE PRETTY GOOD AT FERRETING OUT INFORMATION.
THANKS, MISS LANE. BUT I'LL TAKE MY CHANCES HERE.
THEN... MAYBE YOU CAN BE OUR UNDERWORLD CORRESPONDENT?
DEAL!
YA KNOW, GRUB, THERE'S WORSE DOWN HERE THAN THE WAR-WORLDERS.
YOU THINK I SHOULD TELL 'EM... MY DUTY AS A CORRESPONDENT AN' ALL?
NAH, BLOOD-THIRST IS OUR PROBLEM. BUT IT'S GOOD TO KNOW IF WE CAN'T HANDLE 'IM...
...WE CAN CALL IN SUPERMAN!

ELSEWHERE...
VROOM!
WHAT THE HECK--?!
HEY, BUDDY! GET OUTTA THE WAY!!
WHADDAYA THINK YER--
KRSHWAM!

HA HA HA HA HA HA HA HA HA HA
STATE TROOPERS! CHUCK JOHNSTON CALLING STATE TROOP--
I READ YOU, MR. JOHNSTON. WHAT IS IT?
BIG MONSTER FLIPPED MOON'S RIG... ONE HAND TIED BEHIND ITS BACK!
IT'S BURNIN' FIT TA POP!
DID YOU SAY... "MONSTER"?
YEAH... BIG AS A @#*% HOUSE!
HELLO... WHAT HAVE WE HERE?
THE DUDE'S TEARIN' UP THE WHOLE INTER-STATE!
HE'S HEADING EAST! PLEASE... YOU'VE GOT TO STOP HIM!
NOW, THIS SOUNDS LIKE A JOB FOR THE JUSTICE LEAGUE!

QUICKLY, BOOSTER! THESE MEN REQUIRE MEDICAL ATTENTION!
SO LET'S GET OUT OF THIS INFERNO, MAXIMA--
--BEFORE IT MANAGES TO EAT ITS WAY RIGHT THROUGH MY FORCE FIELD!
BEFORE YOU GO ANY FURTHER... PICK UP AND READ MAN OF STEEL #18-- STILL ON SALE!
DOWN for the COUNT
DAN JURGENS story and art
RICK BURCHETT finished art
WILLIE SCHUBERT letters
RUBEN DIAZ asst. editor
GENE D'ANGELO colors
BRIAN AUGUSTYN editor

I'LL HAVE THIS LITTLE CAMPFIRE SNUFFED OUT SOONER THAN YOU CAN SAY WEENIE ROAST!
LEXOIL

--JUST AS I KNOW WE HAVE RUMORS OF A MONSTER MAN RUNNING LOOSE THAT CAUSED THIS ACCIDENT.
IT'S TOO BAD HE VANISHED INTO THE WOODS OR WE'D TAKE CARE OF HIM!
ANY WORD ON THE PARAMEDICS, BOOSTER?

I HAVE MY WRIST COMMUNICATORS TUNED INTO THE POLICE BAND.
DISPATCH SAYS AMBULANCES WILL BE HERE WITHIN TWO MINUTES!

MY ICE WILL MELT SOME OF THE BLAZE!
TSSSSSSSSS
GOOD MOVE, ICE! THE REST OF US WILL TAKE CARE OF THE VICTIMS!
WE SURE APPRECIATE THE HELP, JUSTICE LEAGUERS! I'M WELL AWARE THAT OHIO IS OUT OF YOUR NORMAL AREA OF JURISDICTION--
WE MUST FIND THE BEAST.
I AGREE, BLOODWYND. EVERYBODY INTO THE BUG AND WE'LL BE ON OUR WAY.
GOSH, IT'S TOO BAD WE HAVE THIS EMERGENCY! I REALLY WANTED TO SEE TODAY'S EPISODE--
"--OF THE CAT GRANT SHOW."
WE'RE COMING TO YOU LIVE FROM ROOSEVELT HIGH WITH AN INCREDIBLE SHOW!
HE IS PERHAPS THE MOST CELEBRATED MAN OF OUR TIME! HE'S BEEN CALLED THE MAN OF TOMORROW--
--THE LAST SON OF KRYPTON AND THE MAN OF STEEL! BUT HE'S MOST APPROPRIATELY KNOWN AS--

"--SUPERMAN!"
YEAH!
YAYYYY
CLAPCLAP
CLAP CLAPCLAP
KEEP YOUR EYES ON THE GROUND, PEOPLE! THE SOONER WE SPOT OUR MONSTER THE BETTER!
HEY, BEETLE, IF IT'S A REALLY COOL MONSTER MAYBE WE SHOULD CAPTURE IT--
--AND TAKE IT ON THE TALK-SHOW CIRCUIT FOR BIG BUCKS!
YOUR SENSE OF THE APPROPRIATE KNOWS NO LIMITS, BOOSTER.
IT TOOK YOU THIS LONG TO REALIZE THAT, BLOODWYND?
UH-OH...LOOKS LIKE WE'VE FOUND OUR MAN'S TRAIL OF CRUMBS!
CHECK OUT THAT PATH OF DESTRUCTION!
THOSE TREES WEREN'T MOWED DOWN BY A COUPLE OF KIDS ON SKATE-BOARDS!
THIS IS TERRIBLE! SUCH POINT-LESS...NEEDLESS DEVASTATION!
LET'S JUST FIND THE SUCKERS AND KICK SOME BUTT!
4

I CAN'T THANK YOU ENOUGH FOR JOINING US HERE, SUPERMAN. INTERVIEWS WITH YOU ARE A TRUE RARITY!
I'VE ALWAYS FELT THAT IF AMERICANS ARE TO TRUST US, THEY HAVE TO KNOW US, MS. GRANT.
AND WITHOUT YOUR TRUST WE ARE NOT EFFECTIVE.
YOU'VE EXHIBITED PSYCHIC POWERS BEFORE, BLOODWYND. ANY CHANCE YOU CAN SCAN AHEAD AND TAP INTO THIS GUY'S MIND?
IT WILL BE DIFFICULT--
--BUT I CAN TRY.
AS WILL I--
I GUESS TWO PSYCHIC MINDS ARE BETTER THAN ONE.
NUTS! I WANTED BLOODWYND TO GO IT ALONE SO I'D HAVE A CHANCE TO GAUGE HIS POWERS!
IT'S THE ONLY WAY I CAN GET INFO ON THE GUY!
THE WAY HE SHIELDS HIS ABILITIES HE'LL PROBABLY LET MAXIMA MAKE FIRST CONTACT EVEN IF--
YES!
I'VE FOUND THE CREATURE!
HE'S HATE--
--DEATH AND BLOOD LUST PERSONIFIED!
NOTHING MORE.

AGREED. SOME OF YOUR COLLEAGUES, LIKE BOOSTER GOLD, ELONGATED MAN AND WONDER WOMAN--
--HAVE LED VERY PUBLIC LIVES! BUT WE DON'T KNOW NEARLY AS MUCH ABOUT YOU!
AS LEADER OF THE JLA PERHAPS YOU CAN GIVE US THE INSIDE STORY ON YOU AND YOUR PALS.
LET ME CORRECT YOU ON THAT POINT, MS. GRANT.
crunch
crunch crunch
BRAMMMMM
sniff
sniff

IT'S UNFAIR TO THE OTHERS TO PAINT ME AS THE LEADER OF THE JUSTICE LEAGUE.
WE'RE A GROUP OF PEOPLE WHO HAVE GOTTEN TOGETHER TO DO A JOB ONLY WE CAN DO.
EVERYBODY IN THE GROUP HAS A SAY ON ISSUES... AND A VOTE AS WELL.
GLURTCH
CRACKK
HA HA HAA!
?

SURELY YOU'VE BEEN A GREATER INFLUENCE THAN THAT!
RESPECTED OBSERVERS SUGGEST YOU'VE PROVIDED A QUALITY OF STRENGTH AND FOCUS THE LEAGUE PREVIOUSLY LACKED.
THEY WERE A TALENTED, DEDICATED BUNCH LONG BEFORE I JOINED, MS. GRANT.
I'M PROUD TO BE IN THEIR RANKS.
CAMERA ONE
CAMERA TWO
YO, BEETLE! RADAR SHOWS A PROJECTILE COMING--
BLIP B
EVERYBODY ASSUME CRASH POSITIONS!
SHRAK
OUR HYDRAULICS ARE SHREDDED! WE'RE GOING DOWN!
I'M GONNA FIND THE GUY WHO WHACKED US AND SEW HIS EYELIDS SHUT!
BETTER GIVE THE NON-FLYERS A HAND FIRST!
AND YOU BETTER MAKE IT QUICK!

DON'T BE SO MODEST, SUPERMAN! THE WHOLE CITY KNOWS YOU AND GUY GARDNER EXCHANGED BLOWS A FEW WEEKS AGO!
CARE TO COMMENT ON THE CONFLICT BETWEEN YOU TWO?
MISTER GARDNER AND I ARE COLLEAGUES. THOUGH IT'S TRUE OUR RELATIONSHIP IS SOMEWHAT... ICY...
...WE DO MANAGE TO GET THE JOB DONE.
FEELS GOOD TO GET MY HANDS ON YOU AGAIN, BABE!
WHAT'S SUPERDOOF GOT THAT I DON'T?
CLASS.
GOTCHA!
JUST SAVE THE SHIP'S WRECKAGE BEFORE IT KILLS SOMEONE!
DONE. WITH MY COMMAND OVER METALS I CAN EASILY LOWER IT TO THE GROUND.
WHO'S BEHIND THIS? WEAPONS MASTER? DESPERO?
NEITHER. AND I SENSE THE TRUE CREATURE RESPONSIBLE IS NEARBY.

OKAY, IT'S TIME TO LET THESE STUDENTS ASK SOME QUESTIONS. YOUR NAME IS...
KENNY. I WAS WONDERIN' WHAT YOU SUPER-HEROES DO WHEN YOU'RE NOT BASHING BAD GUYS?
I MEAN, DO YOU GET TOGETHER AND PARTY ALL THE TIME OR WHAT?
KOULA BEER
HFF!
MMM...
MMMAHAHAHAA!

WE JUSTICE LEAGUERS HAVE A VARIETY OF INTERESTS JUST AS YOU AND YOUR FRIENDS DO.
BLUE BEETLE IS AN INVENTOR WHO ENJOYS SPENDING HIS FREE TIME IN THE LAB, ICE LIKES TO TRAVEL--
--MAXIMA IS BUSY ADJUSTING TO EARTH LIFE WHILE GUY GARDNER TENDS TO BE A LITTLE MORE... CREATIVE IN HIS FREE TIME.
AN EXPLOSION BEYOND THAT GROVE OF TREES! I THINK THERE'S A LEXOIL REFINERY OVER THERE!
LIKE, DO YOU THINK THE GUY WE'RE LOOKING FOR CAUSED IT?
ONLY ONE WAY TO FIND OUT! BOMBS AWAY!
WAIT, GUY! LET'S GET A LITTLE ORGANIZED HERE!
TOO LATE, BUG! ONE SOON-TO-BE CORPSE DEAD AHEAD!
TAKE YOUR PICK! BURIAL OR CREMATION?
LEX OHIO
HURF?

SPEAKING OF GUY GARDNER, WHY WON'T YOU GUYS LET HIM BE A GREEN LANTERN ANYMORE?
WHY DID YOU FIRE HIM?
I CAN ASSURE YOU THAT WE HAD NO SAY REGARDING GUY'S STATUS IN THE GREEN LANTERN CORPS.
IT WAS THEIR CALL ALL THE WAY.
Huh?
BAA!
HA HAA!
Uhn! SO FAST I DIDN'T EVEN SEE HIM MO--
KRAKOWW
LexOil
OHIO FACILITY
SPLAKK!

THIS INTERVIEW IS TAKING PLACE IN METROPOLIS FOR THE BENEFIT OF HIGH SCHOOLS NATIONWIDE. I WANT YOU KIDS TO PAY ATTENTION!
Next Assignment
IF YOU ASK ME, IT'S A YAWNER, MITCH.
NO KIDDIN'. THEY SHOULDA TALKED TO GUY GARDNER--
"--IF THEY WANTED TO TALK TO SOMEONE WITH THEIR HEAD SCREWED ON STRAIGHT!"
HAHAHAHAA!
LET GO OF HIM, YOU MONSTER!
EEF!
THAT FREAKIN' THING'S POWERFUL! NEVER HAD A CHANCE TO FIGHT BACK!
CAN'T BREATHE...
CAN'T... ...EVEN... ...THINK...

YEAH, I WAS KINDA WONDERIN' ABOUT FIRE.
DOES SHE SCORE AS HIGH ON THE BABE-O-METER AS IT SEEMS?
YUK! NICE QUESTION, SPUD!
FIRE IS GOOD AT HER JOB AND A TERRIFIC PERSON. NEXT QUESTION.
I CAN'T BELIEVE IT! NO MATTER HOW MUCH I THROW AT THIS GOON, IT DOESN'T FAZE HIM!
THEN I WILL DEAL WITH HIM.
OVER THE YEARS MANY SOULS HAVE PASSED TO THE AFTERLIFE HERE.
THIS AREA IS RICH WITH THE PLASMA ENERGIES OF THE SPIRITS OF THE DEAD.
I CALL ON THOSE SPIRITS--
--TO GRANT ME THE POWER NECESSARY TO DEFEAT THIS MONSTROSITY!
BAMM
GODS! WHAT MANNER OF CREATURE CAN WITHSTAND SUCH A BLOW?

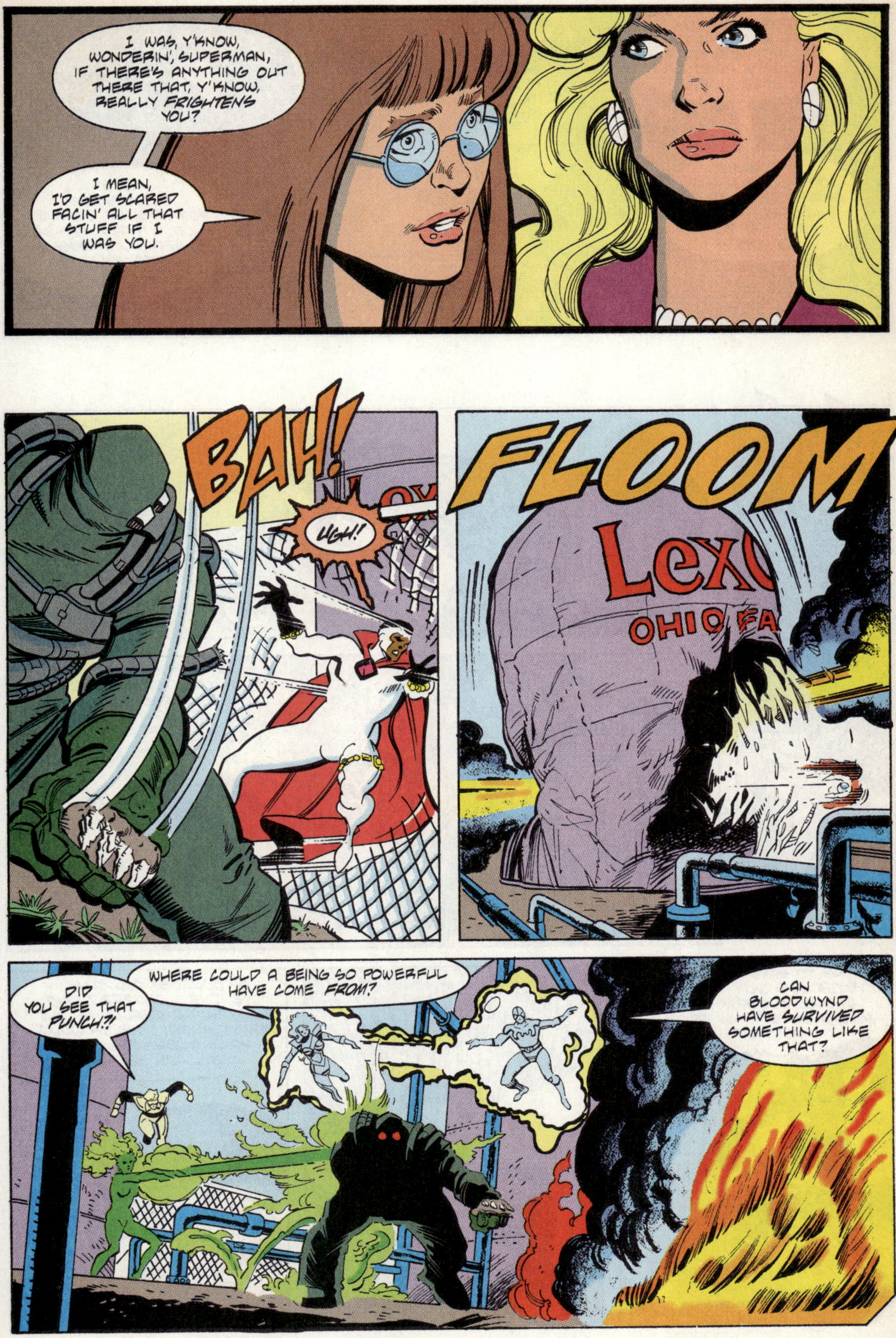
I WAS, Y'KNOW, WONDERIN', SUPERMAN, IF THERE'S ANYTHING OUT THERE THAT, Y'KNOW, REALLY FRIGHTENS YOU?
I MEAN, I'D GET SCARED FACIN' ALL THAT STUFF IF I WAS YOU.
BAH!
UGH!
FLOOM
LEXO
OHIO FA
DID YOU SEE THAT PUNCH?!
WHERE COULD A BEING SO POWERFUL HAVE COME FROM?
CAN BLOODWYND HAVE SURVIVED SOMETHING LIKE THAT?

GOOD QUESTION, MISS. SEE, ONE WAY OR ANOTHER, FEAR IS ALWAYS PART OF THE JOB.
I'M AFRAID OF FAILURE AND AFRAID OF HURTING INNOCENT PEOPLE AND, TO BE CANDID--
--I'VE BEEN AFRAID FOR MYSELF. I HAVE ENCOUNTERED THINGS POWERFUL ENOUGH TO KILL ME.
HEH.
HA!
OH!
YOU GUYS TAKE CARE OF THE STEROID CASE! I'LL GET BLOODWYND OUT OF THAT INFERNO!
WEIRD! HERE I AM TRYING TO SAVE THE MOST MYSTERIOUS GUY IN THE LEAGUE! BLOODWYND IS HIDING SOMETHING FROM US THAT--
THERE! BUT THAT'S NOT--
OF COURSE! ALL THIS TIME I'VE WONDERED WHO BLOODWYND REALLY IS AND NOW I KNOW! I NEVER WOULD HAVE GUESSED IT IN A MILLION YEARS--
--BUT BLOODWYND IS REALLY--

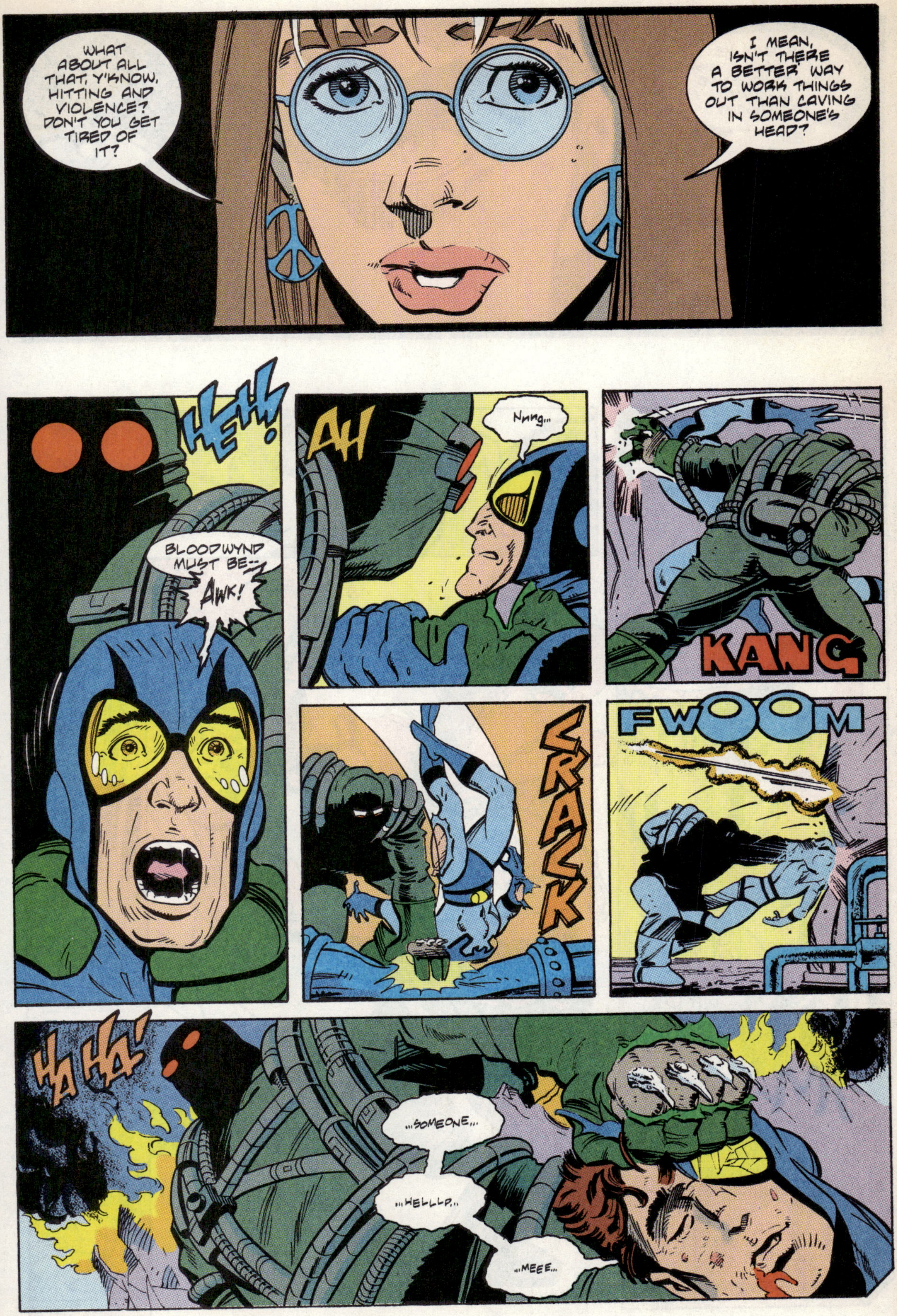

WHAT ABOUT ALL THAT, Y'KNOW, HITTING AND VIOLENCE? DON'T YOU GET TIRED OF IT?
I MEAN, ISN'T THERE A BETTER WAY TO WORK THINGS OUT THAN CAVING IN SOMEONE'S HEAD?
HEH!
BLOODWYND MUST BE--
AWK!
AH
Nnng...
KANG
CRACK
FWOOM
HA HA!
...SOMEONE...
...HELLLP...
...MEEE...

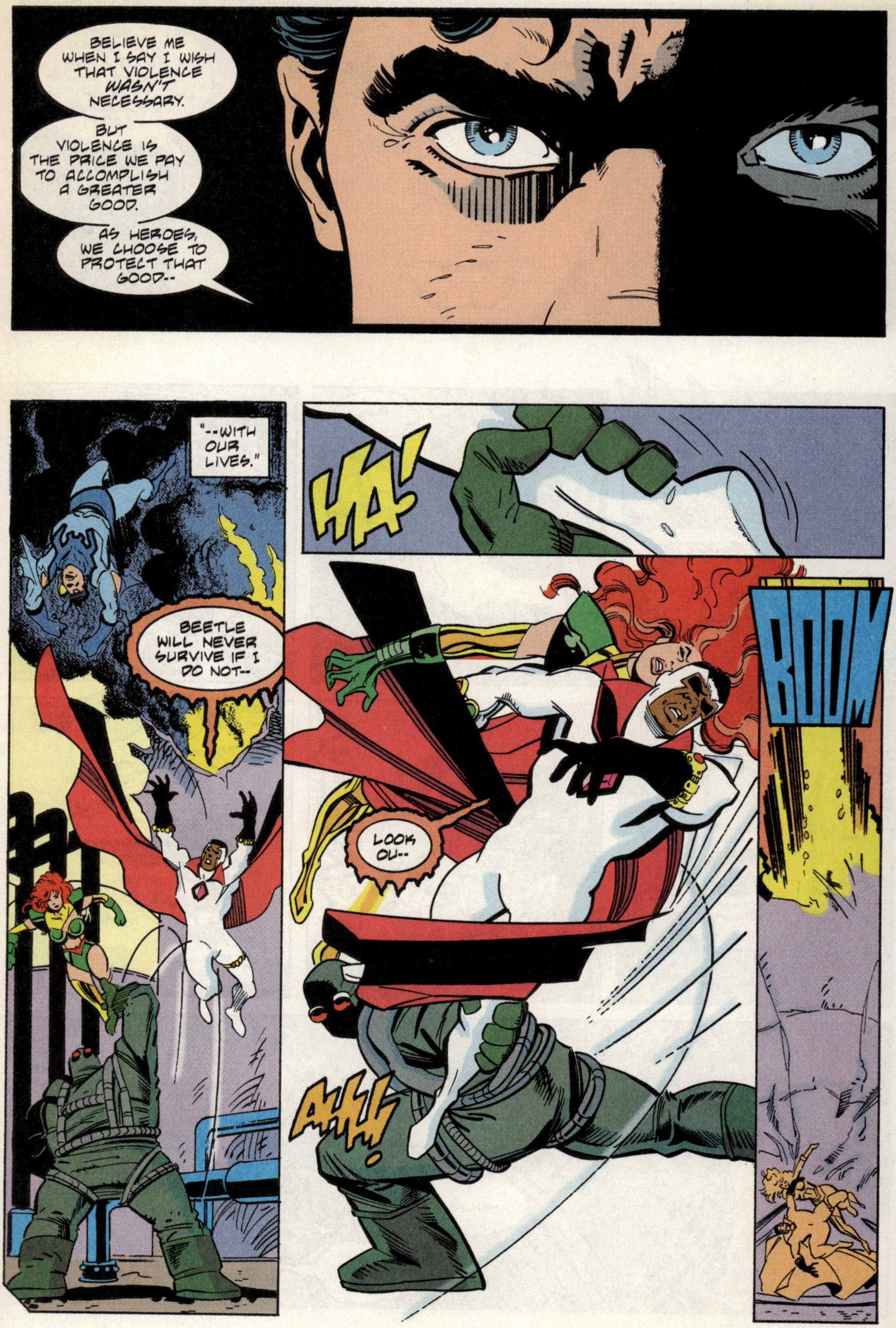
BELIEVE ME WHEN I SAY I WISH THAT VIOLENCE WASN'T NECESSARY.
BUT VIOLENCE IS THE PRICE WE PAY TO ACCOMPLISH A GREATER GOOD.
AS HEROES, WE CHOOSE TO PROTECT THAT GOOD--
"--WITH OUR LIVES."
BEETLE WILL NEVER SURVIVE IF I DO NOT--
HA!
BOOM
LOOK OU--
AHH!

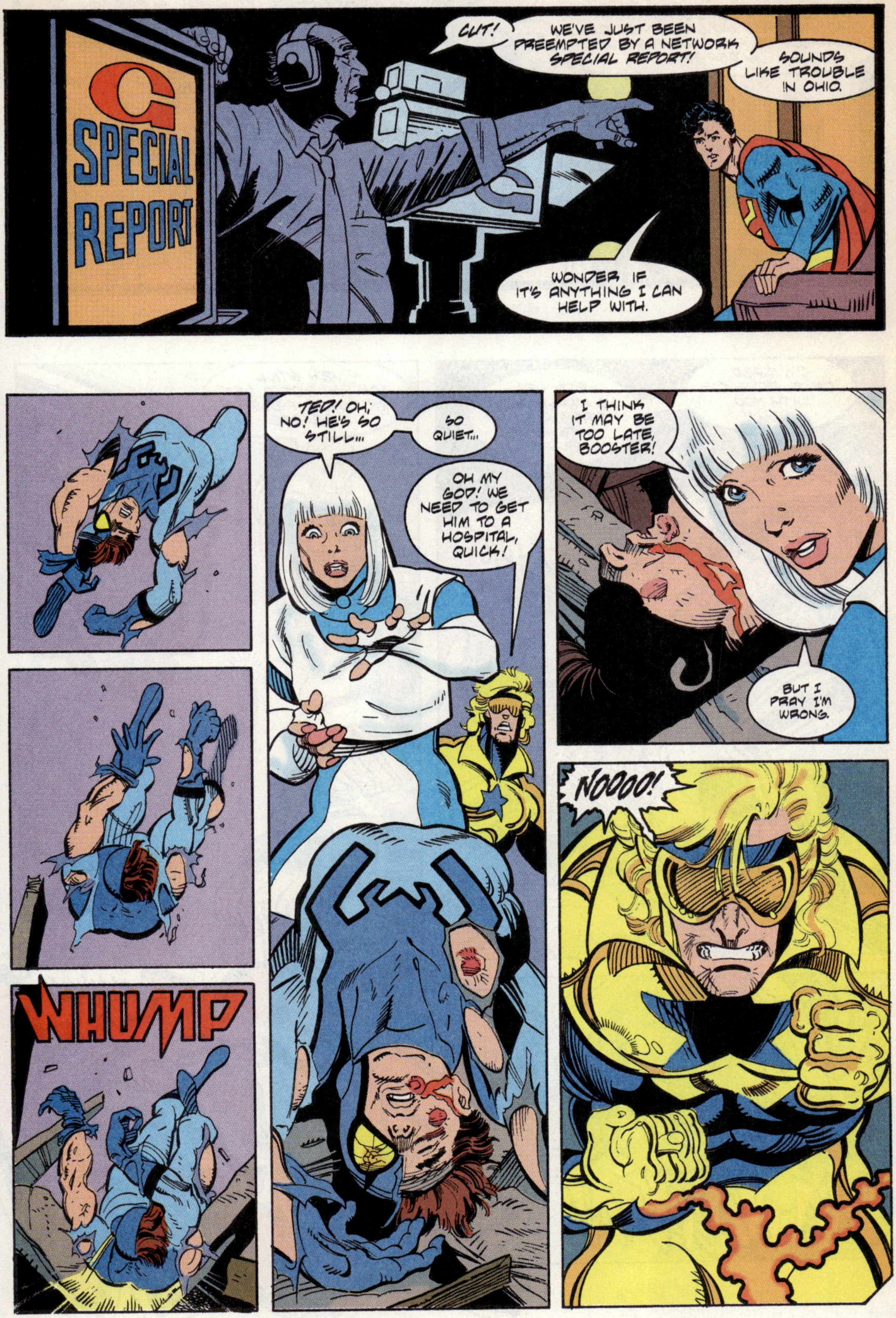
SPECIAL REPORT
CUT!
WE'VE JUST BEEN PREEMPTED BY A NETWORK SPECIAL REPORT!
SOUNDS LIKE TROUBLE IN OHIO.
WONDER IF IT'S ANYTHING I CAN HELP WITH.
WHUMP
TED! OH, NO! HE'S SO STILL...
SO QUIET...
OH MY GOD! WE NEED TO GET HIM TO A HOSPITAL, QUICK!
I THINK IT MAY BE TOO LATE, BOOSTER!
BUT I PRAY I'M WRONG.
NOOOO!

--HAVE REPORTS OF THE JUSTICE LEAGUE BATTLING A HEINOUS MONSTER AT A LEXOIL REFINERY IN OHIO. REPORTS INDICATE THE LEAGUE IS UNABLE TO STOP HIS DESTRUCTIVE STAMPEDE.
SUPERMAN...
I HAVE TO GO.
I'M TIRED OF PLAYING TAG WITH YOU, UGLY!
LET'S SEE, YOU WALK AWAY FROM A FULL-INTENSITY BLAST!
BRAK!
HE'S STILL COMING! GOTTA GET MY FORCE FIELD UP BEFORE--
AAAUHH!
POW
HA HA HAA!

HOW COULD ONE MAN STAND AGAINST THE WHOLE LEAGUE?
SOON AS I'M UP OVER THE HORIZON, I'LL BE ABLE TO CHECK THINGS OUT WITH MY TELESCOPIC VISION!
MAN, I JUST BARELY RAISED MY FORCE FIELD IN TIME.
THE FORCE OF HIS BLOW IS OVERPOWERING MY FLIGHT RING. IF HE CAN DO THIS TO ME--
--IMAGINE WHAT HE DID TO BEETLE.
WONDER IF THAT RUBBER-SUITED POWERHOUSE IS BIOLOGICAL--
--OR SOME KIND OF DOOMSDAY MACHINE!
THAT'S FAR ENOUGH, BOOSTER.
YOU!

I HEAR YOU PEOPLE ARE HAVING SOME TROUBLE.
"TROUBLE" ISN'T THE WORD, SUPERMAN!
I'M TELLING YOU RIGHT NOW--
--IT'S LIKE DOOMSDAY IS HERE!

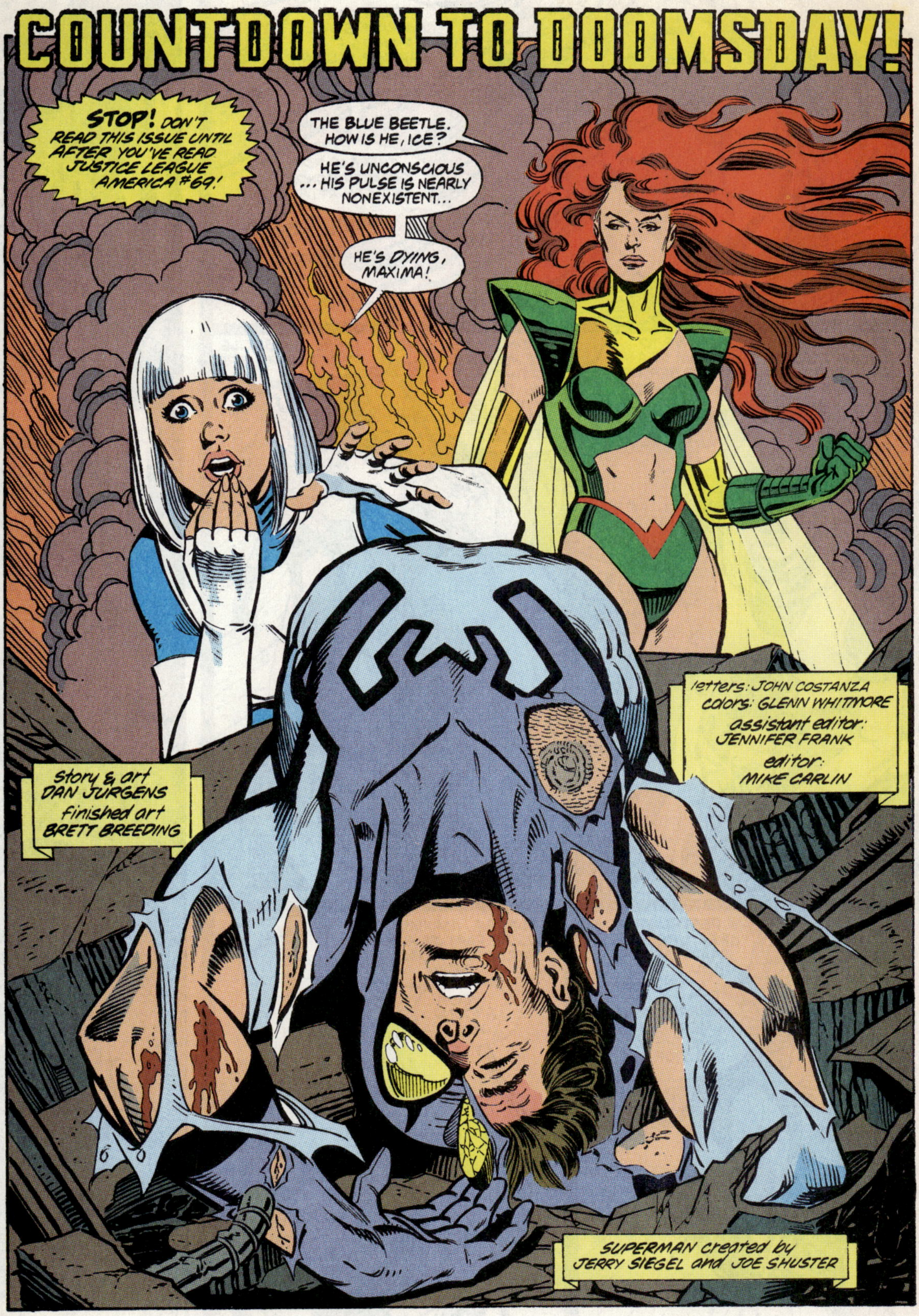
COUNTDOWN TO DOOMSDAY!
STOP! DON'T READ THIS ISSUE UNTIL AFTER YOU'VE READ JUSTICE LEAGUE AMERICA #69!
THE BLUE BEETLE. HOW IS HE, ICE?
HE'S UNCONSCIOUS ... HIS PULSE IS NEARLY NONEXISTENT...
HE'S DYING, MAXIMA!
story & art DAN JURGENS finished art BRETT BREEDING
letters: JOHN COSTANZA colors: GLENN WHITMORE assistant editor: JENNIFER FRANK editor: MIKE CARLIN
SUPERMAN created by JERRY SIEGEL and JOE SHUSTER

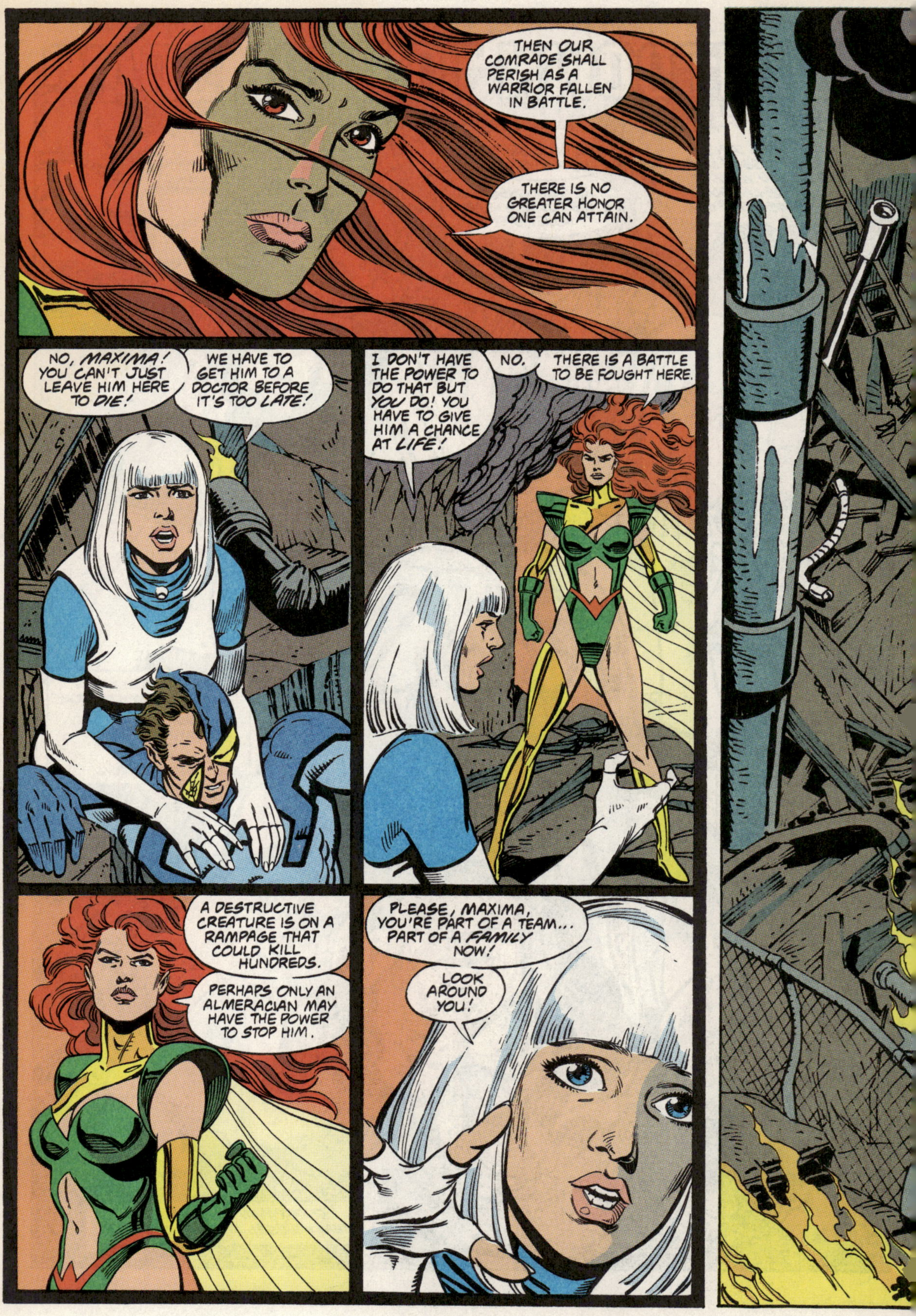
THEN OUR COMRADE SHALL PERISH AS A WARRIOR FALLEN IN BATTLE.
THERE IS NO GREATER HONOR ONE CAN ATTAIN.
NO, MAXIMA! YOU CAN'T JUST LEAVE HIM HERE TO DIE!
WE HAVE TO GET HIM TO A DOCTOR BEFORE IT'S TOO LATE!
I DON'T HAVE THE POWER TO DO THAT BUT YOU DO! YOU HAVE TO GIVE HIM A CHANCE AT LIFE!
NO.
THERE IS A BATTLE TO BE FOUGHT HERE.
A DESTRUCTIVE CREATURE IS ON A RAMPAGE THAT COULD KILL HUNDREDS.
PERHAPS ONLY AN ALMERACIAN MAY HAVE THE POWER TO STOP HIM.
PLEASE, MAXIMA, YOU'RE PART OF A TEAM... PART OF A FAMILY NOW!
LOOK AROUND YOU!

THIS CREEP IS TOUGH... THAT MUCH IS OBVIOUS.
BUT RIGHT NOW ONLY *YOU* CAN SAVE TED KORD'S LIFE.
LET ME GO AFTER THE MONSTER UNTIL YOU GET BACK.
I AM A WARRIOR. IT IS AGAINST MY NATURE TO LET OTHERS DO MY WORK.
YET YOU AIDED ME IN MY STRUGGLES AGAINST STARBREAKER.
I OWE YOU *MY* AID IN RETURN.
I HOPE YOU CAN SURVIVE THE MONSTER'S WRATH UNTIL MY RETURN.
FOR NOW... BLUE BEETLE SHALL RECEIVE HIS MEDICAL CARE.

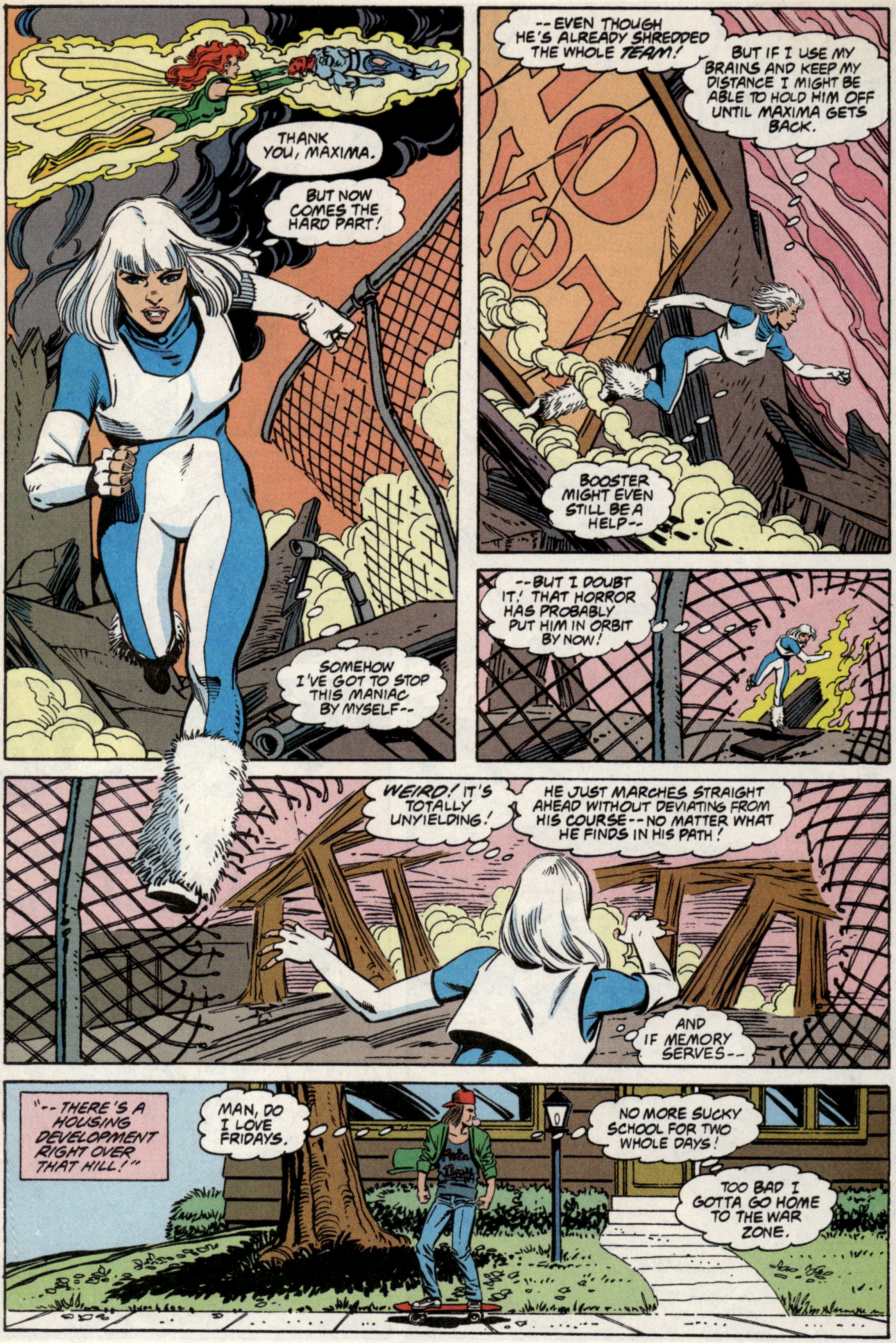
THANK YOU, MAXIMA.
BUT NOW COMES THE HARD PART!
SOMEHOW I'VE GOT TO STOP THIS MANIAC BY MYSELF--
--EVEN THOUGH HE'S ALREADY SHREDDED THE WHOLE TEAM!
BUT IF I USE MY BRAINS AND KEEP MY DISTANCE I MIGHT BE ABLE TO HOLD HIM OFF UNTIL MAXIMA GETS BACK.
BOOSTER MIGHT EVEN STILL BE A HELP--
--BUT I DOUBT IT! THAT HORROR HAS PROBABLY PUT HIM IN ORBIT BY NOW!
WEIRD! IT'S TOTALLY UNYIELDING!
HE JUST MARCHES STRAIGHT AHEAD WITHOUT DEVIATING FROM HIS COURSE--NO MATTER WHAT HE FINDS IN HIS PATH!
AND IF MEMORY SERVES--
"--THERE'S A HOUSING DEVELOPMENT RIGHT OVER THAT HILL!"
MAN, DO I LOVE FRIDAYS.
NO MORE SUCKY SCHOOL FOR TWO WHOLE DAYS!
TOO BAD I GOTTA GO HOME TO THE WAR ZONE.

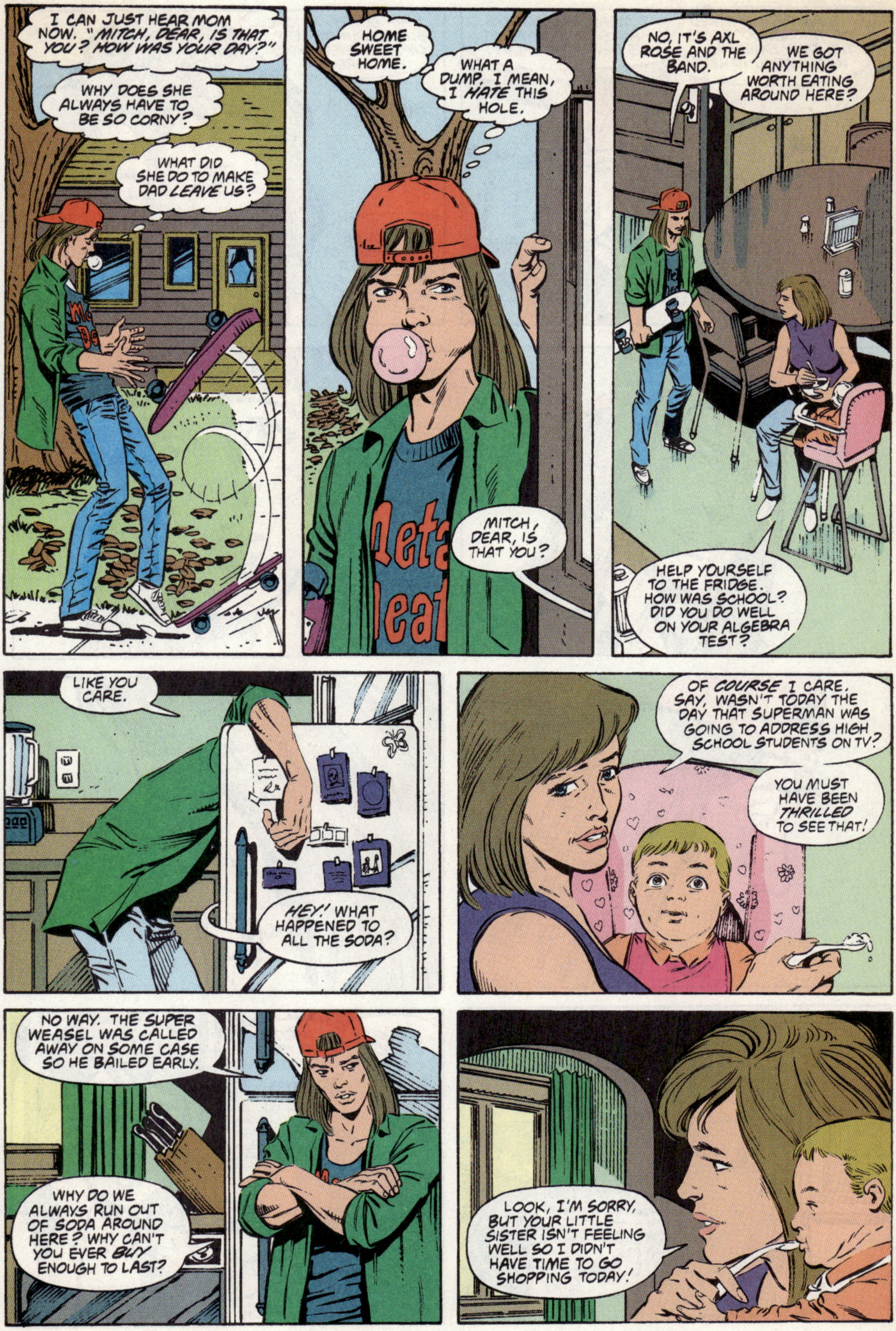
I CAN JUST HEAR MOM NOW. "MITCH, DEAR, IS THAT YOU? HOW WAS YOUR DAY?"
WHY DOES SHE ALWAYS HAVE TO BE SO CORNY?
WHAT DID SHE DO TO MAKE DAD LEAVE US?
HOME SWEET HOME.
WHAT A DUMP. I MEAN, I HATE THIS HOLE.
MITCH, DEAR, IS THAT YOU?
NO, IT'S AXL ROSE AND THE BAND.
WE GOT ANYTHING WORTH EATING AROUND HERE?
HELP YOURSELF TO THE FRIDGE. HOW WAS SCHOOL? DID YOU DO WELL ON YOUR ALGEBRA TEST?
LIKE YOU CARE.
HEY! WHAT HAPPENED TO ALL THE SODA?
OF COURSE I CARE. SAY, WASN'T TODAY THE DAY THAT SUPERMAN WAS GOING TO ADDRESS HIGH SCHOOL STUDENTS ON TV?
YOU MUST HAVE BEEN THRILLED TO SEE THAT!
NO WAY. THE SUPER WEASEL WAS CALLED AWAY ON SOME CASE SO HE BAILED EARLY.
WHY DO WE ALWAYS RUN OUT OF SODA AROUND HERE? WHY CAN'T YOU EVER BUY ENOUGH TO LAST?
LOOK, I'M SORRY, BUT YOUR LITTLE SISTER ISN'T FEELING WELL SO I DIDN'T HAVE TIME TO GO SHOPPING TODAY!

I AM REALLY TIRED OF THAT BABY BEING THE ONLY ONE WHO RATES AROUND HERE!
I MEAN, DAD ALWAYS HAS SODA FOR ME AT HIS NEW APARTMENT!
I AM SORRY, MITCHELL, BUT I CANNOT KEEP UP WITH EVERYTHING HERE!
THIS HOUSE ISN'T PERFECT AND NEITHER AM I BUT WE DO THE BEST WE CAN!
JEEZ.
NO WONDER DAD LEFT AND WANTS A DIVORCE.
GOIN' OVER TO AARON'S.
SEE YOU.
COOO! COOO!
WAIT! DID YOU HEAR THAT--

SKRASSH
OH, MAN!
BECKY...! THE GLASS!
WHUMP
MITCHELL, I WANT YOU TO CALL 911! HURRY!
OH, GOD... OUTSIDE IN THE DRIVE-WAY...
OUR CAR!
CHECK IT OUT!
THAT DUDE DID ALL THIS-- WITH ONE HAND TIED BEHIND HIS BACK!?

HRRR?
THAT'S THE GUY, SUPERMAN. HE'S THE ONE WHO TOOK THE JUSTICE LEAGUE APART AT THE SEAMS.
WHAT WAS IT YOU CALLED HIM, BOOSTER?
OH, YEAH...
DOOMSDAY.
KRAKADOOOM
WOW! THAT PUNCH LOOKED LIKE IT COULD HAVE CAVED IN A MOUNTAIN AND SUPERMAN TOOK IT!
BIG DEAL. THE SPUD WAS TOO SLOW AND STUPID TO DUCK.

RRRR RAAGH!
OOOOF!
KKRASSH
GET DOWN!
CRASH!
KTTTSH
SKKASSH!
UNBELIEVABLE!
BUDDADOOOM
I DON'T KNOW IF I'VE EVER BEEN HIT THAT HARD!
BROW!

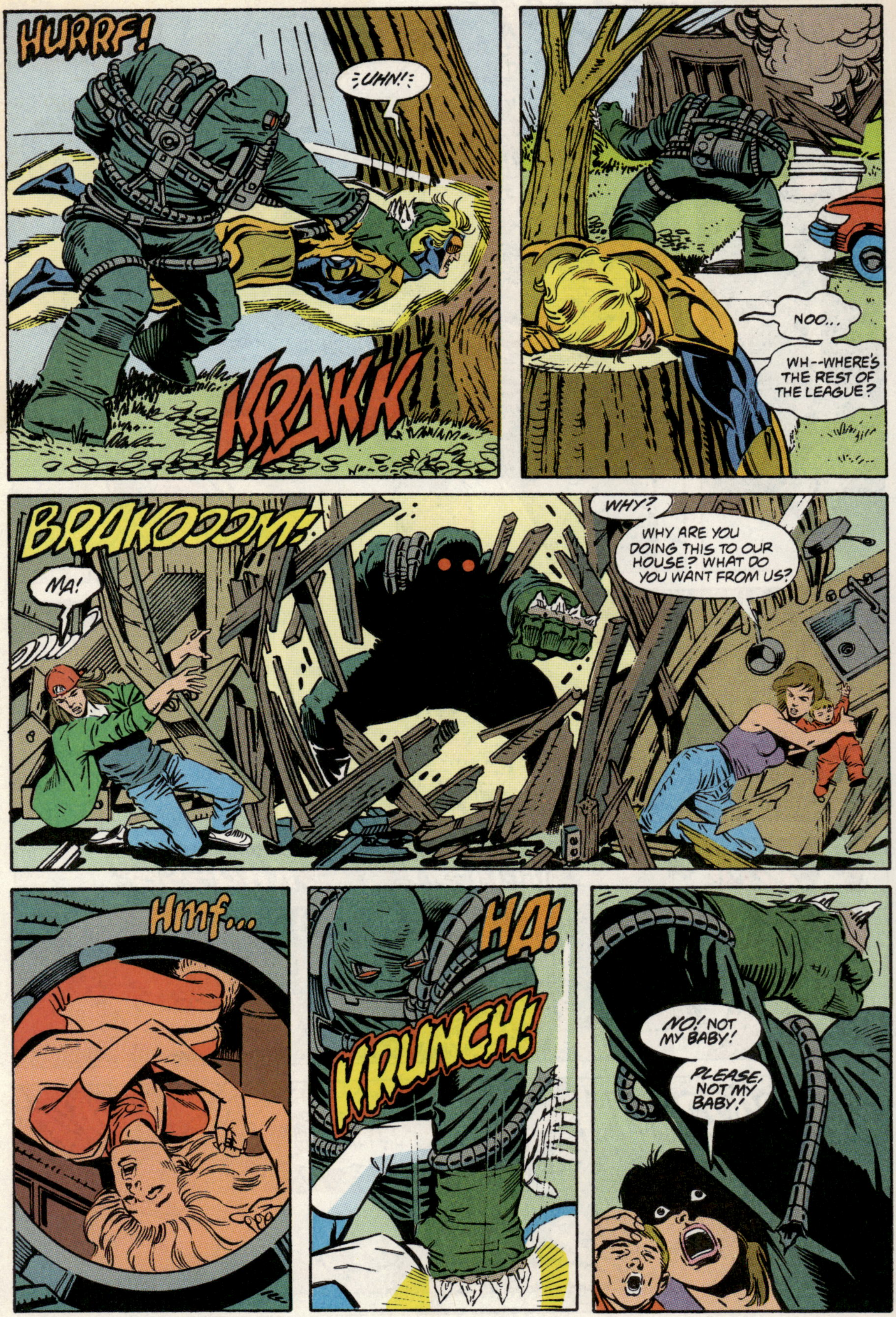
HURRF!
UHN!
KRAKK
NOO...
WH--WHERE'S THE REST OF THE LEAGUE?
BRAKOOOM!
MA!
WHY?
WHY ARE YOU DOING THIS TO OUR HOUSE? WHAT DO YOU WANT FROM US?
Hmf...
HA!
KRUNCH!
NO! NOT MY BABY!
PLEASE, NOT MY BABY!

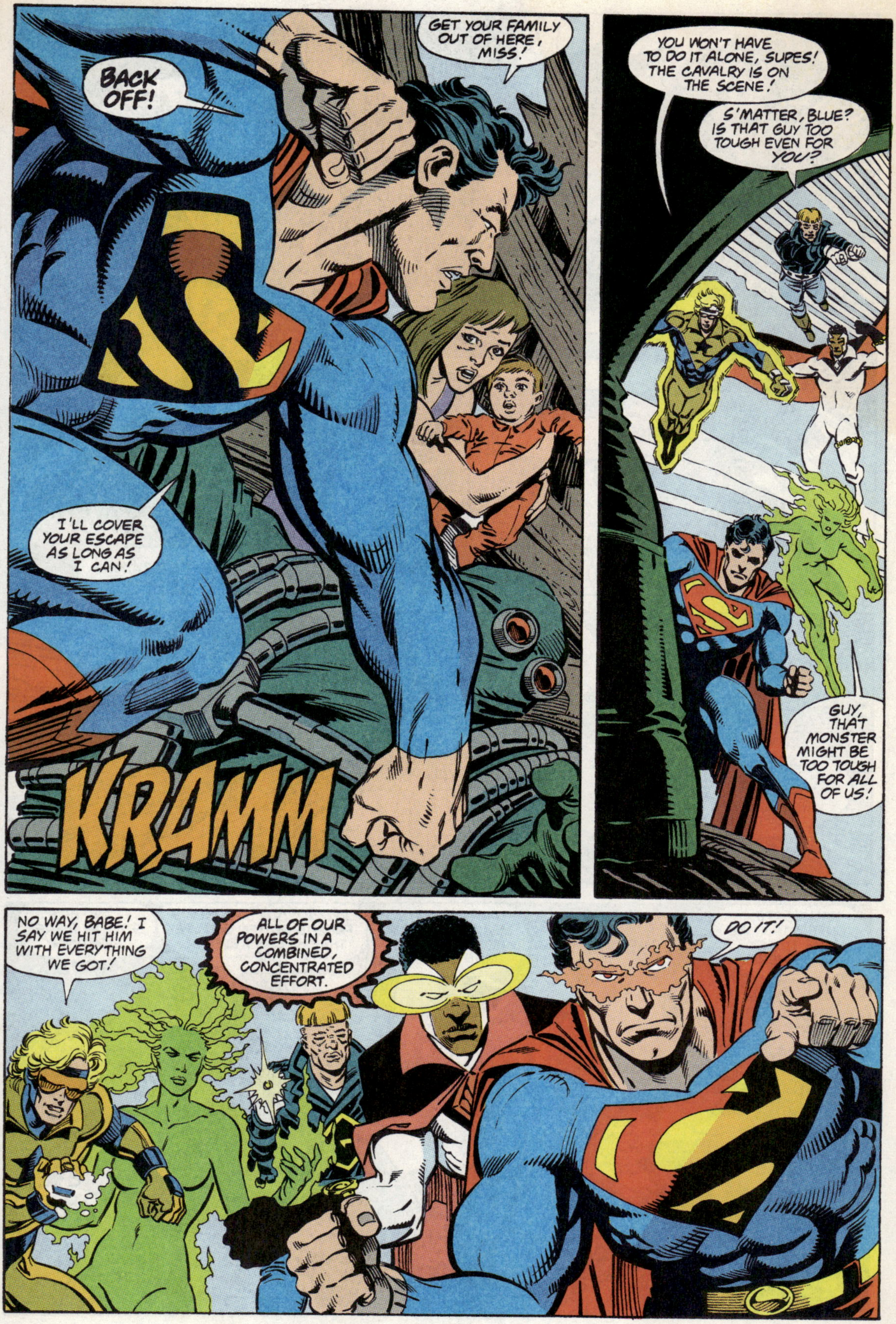

BACK OFF!
GET YOUR FAMILY OUT OF HERE, MISS!
I'LL COVER YOUR ESCAPE AS LONG AS I CAN!
KRAMM
YOU WON'T HAVE TO DO IT ALONE, SUPES! THE CAVALRY IS ON THE SCENE!
S'MATTER, BLUE? IS THAT GUY TOO TOUGH EVEN FOR YOU?
GUY, THAT MONSTER MIGHT BE TOO TOUGH FOR ALL OF US!
NO WAY, BABE! I SAY WE HIT HIM WITH EVERYTHING WE GOT!
ALL OF OUR POWERS IN A COMBINED, CONCENTRATED EFFORT.
DO IT!

HMF...
POUR IT ON!
CAN'T SEE, SO SOMEBODY BETTER POINT MY RING AT THE SUCKER'S UGLY FACE!
DONE.
GIVE IT EVERYTHING YOU GOT--
MMMMM...
--AND THIS DUDE WILL FIND OUT WHAT KIND OF TROUBLE HE BUYS WHEN HE TAKES ON THE JLA!
YOU HEARD SUPERMAN! WE HAVE TO GET OUT OF HERE!
AND WE CAN'T LEAVE THIS POOR WOMAN IN THE LINE OF FIRE!
WOW! GUY GARDNER'S FACE IS BEATEN SO PUFFY THAT HE CAN'T EVEN SEE--
--AND HE'S STILL IN THERE FIGHTING!
COOL!

AMAZING! I CAN'T EVEN SEE HIM ANYMORE BUT I THINK HE'S STILL STANDING!
DON'T STAND THERE BLABBIN', BLUE! JUST TURN UP THE JUICE!
GETTING TIRED...
IGNORE IT, FIRE! JUST KEEP PUSHIN'!
NO! MY FLAME IS TOTALLY SPENT!
CAN'T GO... ANYMORE!
SAME HERE!
MY POWER CELLS ARE SHOT --DRAINED!
AND WITHOUT MY SUIT'S POWERS I'M ABOUT AS POWERFUL AS PEE WEE HERMAN!
OKAY, LET'S GIVE IT A REST! AFTER ALL THIS--
--THERE'S NO WAY DOOMSDAY CAN STILL BE STANDIN'!

I THINK YOU MAY BE MISTAKEN.
RRRRRRRR
RRARGH!
INCREDIBLE! HE TOOK ALL THAT ENERGY WITHOUT MISSING A BEAT!
FIRE, YOU, GUY GARDNER AND BOOSTER ARE NO LONGER ABLE TO DEFEND YOURSELVES.
HEY!
I SUGGEST YOU ALL WITHDRAW AND LET SUPERMAN AND ME HANDLE THIS MON-STROSITY.
NO TIME, GANG! DOOMSDAY IS COMIN'!

LOOK OUT!
HIS SPEED--!
SWAKK!
CRUNCH
WHERE IS HE-- UHH!

WITH... POWER CELLS GONE I'M IN DEEP TROUBLE!
FORCE FIELD-- GONE!
RRRAW!
HA!
HE'S... FASTER THAN... FLASH! CAN'T GET...
SHRIIPP
RRRR
NO!
HA!
SLAMM
WHAT IS THAT-- THING?
CAN'T FIGHT... BUT I CAN AT LEAST HELP YOU!
BUT WHAT ABOUT-- THAT? WHAT IS IT?
MAYBE A ROBOT... OR SOMETHING FROM ANOTHER DIMENSION! WHATEVER--
"--IT'S DEADLY!"
CAN'T SEE, BUT I CAN HEAR THE BONES BREAKIN'!
GOTTA POINT... BLAST... AND HOPE!

HFFF?
HA!
UHN!
I SHALL BLAST THIS HORROR--
TWO-FRONT OFFENSIVE NOW!
¡UGH!
EERK!
MY EYE BEAMS!
NOOOOO!
JEEZ! BAD ENOUGH THE MONSTER IS TRASHING US! THE LEAGUE DOESN'T HAVE TO HELP!

OH, GOD... IF THAT BLAZE HAS GOTTEN TO ONE OF THE GAS LINES IN THE HOUSE...
BWHOOOOOOM!
CRIPES!
TORA? TORA? ARRRR!
THUD

WHERE... WHERE IS EVERYBODY?
MA? ARE YOU OKAY?
MA?!
SO-- :KAFF: MUCH SMOKE...
CAN'T... BREATHE. CAN'T...
HAHAHA
NO.
NO WAY--
--IS THAT MANIAC ESCAPING ME!

MA!
OH, MAN, SHE MUST BE HURT OR SOMETHING!
THE FIRE'S SURROUNDED US! I'LL NEVER GET HER OUT OF HERE!
NEED HELP...BUT THE JLA LOOKS WAY OUT OF IT OR WORSE!
"FIRE--
"ICE--
"--BOOSTER GOLD...THE BLOOD GUY--
"--EVEN GUY GARDNER!"
THE ONLY ONE WHO CAN HELP US--
--HAS ALREADY BUGGED OUT AFTER THAT KILLING MACHINE!
SUPERMAN!

I DON'T KNOW WHO OR WHAT EXACTLY THIS DOOMSDAY THING IS, BUT I'LL BEAT THE ANSWER OUT OF HIM IF I HAVE TO!
HE DOESN'T APPEAR TO HAVE ANY MAGICAL POWERS SO I WON'T HAVE TO WORRY ABOUT THAT!
PLEASE, SUPERMAN! YOU JUST GOTTA HEAR ME!
THAT BOY I SAW! I CAN HEAR HIM CALLING... BUT I HAVE TO IGNORE IT! I'M SO CLOSE!
HELP US! PLEASE!
IF I LET DOOMSDAY GET AWAY THERE'S NO TELLING WHAT DEVASTATION HE'LL BE RESPONSIBLE FOR.
MUCH AS IT PAINS ME-- I HAVE TO STAY WITH HIM AND BLOCK OUT THAT PLEA FOR HELP!
PLEASE, SUPERMAN...
...YOU JUST GOTTA COME BACK
PLEASE!

"THIS IS TOTALLY NUTS-- I COME HOME FROM SCHOOL, GET INTO ANOTHER FIGHT WITH MA...
"...AND THEN ALL AT ONCE, THE FREAKIN' JUSTICE LEAGUE CRASHES DOWN ON US, ALONG WITH SOMEONE CALLED DOOMSDAY!"
"THE HOUSE IS A DISASTER AREA-- FLAMES ARE EVERYWHERE.
"THAT DOOMSDAY GUY DID IT ALL-- AND JUST BOOKED OUT OF HERE, WITH SUPERMAN ON HIS TAIL!*
"CAN'T HARDLY BELIEVE GUY GARDNER GOT SO BUSTED UP!
"THOSE TWO BABES, ICE AND FIRE, MIGHT BE DEAD-- I CAN'T TELL FROM HERE...
"...BUT FROM THE SOUND OF THINGS ABOUT TWENTY FEET AWAY, WHERE OUR FAMILY ROOM USED TO BE...
"...THAT BOOSTER GOLD MUST WISH HE WAS CROAKED!"
*IN SUPERMAN #74!

"I CAN HEAR MY MA, CALLING OUT TO ME, AND I ANSWER HER, BUT I CAN'T HELP HER OR MY BABY SISTER!
"THROUGH THE THICK BLACK SMOKE, I SEE SUPERMAN, UP IN THE SKY--I HEAR THE SOUND OF HIS FISTS ALL OVER DOOMSDAY!
"PLEASE, GOD--LET HIM HEAR MY CRIES FOR HELP!"
UNDER FIRE
"NO ONE ELSE CAN HELP--THE SIRENS ARE TOO FAR OFF--THEY'LL NEVER REACH US IN TIME.
"IT SEEMS LIKE THIS WHOLE DEAL HAS BEEN GOING ON FOR HOURS, 'THOUGH IT'S PROBABLY ONLY BEEN MINUTES!"
-KAFF KAFF- MITCH, IT'S NO USE--I'VE GOT TO DO SOMETHING...!

TOM GRUMMETT - PENCILLER
DOUG HAZLEWOOD - INKER
JERRY ORDWAY - WRITER
ALBERT DE GUZMAN - LETTERER
GLENN WHITMORE - COLORIST
JENNIFER FRANK - ASSISTANT EDITOR
MIKE CARLIN - EDITOR
SUPERMAN CREATED BY SIEGEL & SHUSTER
"SMOKE'S TOO THICK--MA'S GOT TO HOLD TIGHT! I'VE GOT TO YELL LOUDER-- HE'S GOT TO HEAR ME!"
SUPERMAN! PLEASE-- YOU'VE GOT TO HELP US! MY MA'S TRAPPED-- PLEASE!

SUPERMAN-- YOU'RE THE ONLY ONE--HELP US!
KAFF CAN'T GIVE UP-- UGHNN THAT KID AND HIS MOM DIDN'T ASK FOR--
KAFF-KAFF! GUY--IT HURTS--MY RIBS--I CAN'T STAND UP TOO--UHHN.
FOR GUY'S SAKE, I'VE GOT TO GET HIM UP--THAT MONSTER'S ALREADY PUT BLUE BEETLE INTO A COMA.*
*ALSO IN SUPERMAN #74
MY VOICE'S GOING--FROM BREATHIN' THIS SMOKE--BUT I'VE GOTTA KEEP YELLIN'! UNLESS HE'S NOT LISTENING?
NAH, HE'S A HERO--THEY'RE SUPPOSED TO HELP US! MAYBE THAT DOOMSDAY'S BEATING SUPERMAN? WHAT DO I DO THEN?
HELP US, SUPERMAN--PLEASE!

I HEAR YOU, KID-- I JUST THOUGHT ONE OF THE LEAGUERS MIGHT BE ABLE TO--!
DAMN! GARDNER AND THE REST ARE UNCONSCIOUS! I'VE GOT TO GET BACK DOWN THERE!
"THEY'RE SO HIGH UP, BUT IT LOOKS LIKE-THAT'S IT--THAT'S IT--HE'S LOOKING THIS WAY.
"HE SEES ME!"
WHU-UHHHFF!
CWAMM
"SHOOT! THAT SOUNDED LIKE THUNDER! SUPIE'S GOTTA SHAKE THAT DOOMSDAY CRUD BEFORE HE CAN DO US ANY GOOD!"
THIS CREATURE'S FAST, AND STRONG-- BUT IT SEEMS TO LEAP RATHER THAN FLY.
SO LONG AS I CAN HOLD TIGHT--IT'S AT MY MERCY WHERE INERTIA'S INVOLVED!

I'VE GOT TO FORCE THIS THING FAR ENOUGH INTO THE LAKE'S SILT...
CHOOM
"...AND HOPE IT'S SOFT ENOUGH TO KEEP HIM FROM USING IT TO PROPEL HIMSELF FOR A FEW MINUTES!
"I JUST PRAY I'VE STILL GOT TIME TO HELP THAT FAMILY!"
KERACK
OH, MERCY-- DON'T LET THAT BE THE SOUND OF THOSE BEAMS GIVING WAY!
UGHNN! HOPE IS NOT LOST...
...NOT WHILE BLOODWYND STILL STANDS!
I DON'T KNOW WHERE YOU CAME FROM, MISTER-- BUT THANK GOD YOU'RE HERE!

YOU AND YOUR CHILD SHOULD ALSO THANK SUPERMAN.
BLOODWYND-- ARE YOU OKAY? YOU TOOK QUITE A BEATING--
ALL OF THE LEAGUERS DID-- BUT THIS GUY KEEPS SURPRISING ME WITH HIS--RESILIENCY.
S-SUPERMAN-- WHAT ABOUT MY SON, MITCH?
HE'S DOWN THERE WITH THE E.M.S. CREW, MISS.
THEY DID IT! THEY SAVED MY MOM AND MY BABY SISTER!
YOU SHOULD GO TO THE HOSPITAL-- YOU ALL TOOK IN A LOT OF SMOKE.
GUY-- LIE STILL.
HOW 'BOUT IT, FELLA? LET'S HAVE A LOOK AT YOU.
NO. I DESIRE NO MEDICAL TREATMENT.
I WISH TO BE ALONE.
ODD-- BLOODWYND'S TELEPORTED RATHER THAN SEEK TREATMENT--!
DON'T STRUGGLE-- WHAT IS IT?

-KAFF KAFF- DON'T WUSS OUT, BOYSCOUT! PUT THIS DOOMSDAY GUY IN A PINE BOX--
--OR I'LL CRAWL OFFA THIS GURNEY AND KICK BOTH O' YER BUTTS! -KAFF-
I'LL TAKE CARE OF THINGS, GUY-- YOU JUST LET THE DOCTORS HELP YOU!
YOU THERE-- HAVE YOUR LOCAL HOSPITAL CONTACT MAXWELL LORD IN NEW YORK CITY FOR THESE FOLKS' MEDICAL RECORDS!
NOW TO SEE IF THIS "DOOMSDAY" IS STILL WHERE I LEFT HIM!
"QUESTION IS-- HOW DO I RESTRAIN HIM WHEN THE COMBINED FORCE OF THE JUSTICE LEAGUE COULDN'T DO IT?"
HOLY--! THAT'S OUR TARGET DOWN THERE, RALPH!
IT'S COMING UP FAST, BUT OUR WEAPONS SYSTEM'S LOCKED ON! COMMENCING LAUNCH OF HELLFIRES--
SPLASH!

CRASSH
MAYDAY! MAYDAY-- THE MISSILES ARE RUNNING WILD!
GEEZ, DOES THIS MONSTER JUST LIVE TO DESTROY, OR WHAT?
LET'S HELP THIS MISSILE HIT ITS TARGET, WHILE I ATTEND TO THOSE TWO AIRMEN!
KA-CHOOM
GOT THEM!
AND WHILE "MR. DESTRUCTO'S" IN FREE FALL, I'D BETTER ROUND UP THOSE OTHER TWO ERRANT MISSILES!

KIRBY COUNTY POLICE
SOMETHING TORE UP A STRETCH OF PROPERTY OUT ON ROUTE 110, RUSTY!
LOWELL SAID A BUNCH OF FOLK--INCLUDING SOME OF THE JUSTICE LEAGUE --ARE BEING RUSHED TO THE HOSPITAL!
SHOULD I CRANK UP THE CIVIL DEFENSE SIREN, CHIEF? MAYBE WE SHOULD GET EVERYONE INTO THEIR BASEMENTS--
SAY, YOU HEAR THAT? KIND OF A CARTOON SOUND A BOMB MAKES JUST BEFORE IT--
MOTHER OF PEARL!
CRASHHT
KIRBY COUNTY POLICE STATION
UH, CHIEF-- THINK I'M GONNA NEED A BIGGER GUN!
SSSSSSSSS

SAY... THERE'S THAT SOUND AGA--
THE GLOVES ARE OFF, DOOMSDAY! I'M TIRED OF TREADING LIGHTLY!
WHAAACK!
KA-DOOM
HOT DAMN! THOSE GOOD OLD BOYS ARE TEARING UP MAIN STREET!
GET THE GOVERNOR ON THE PHONE!
UGNHH
IS--IT POSSIBLE THAT THIS GUY'S GETTING STRONGER?

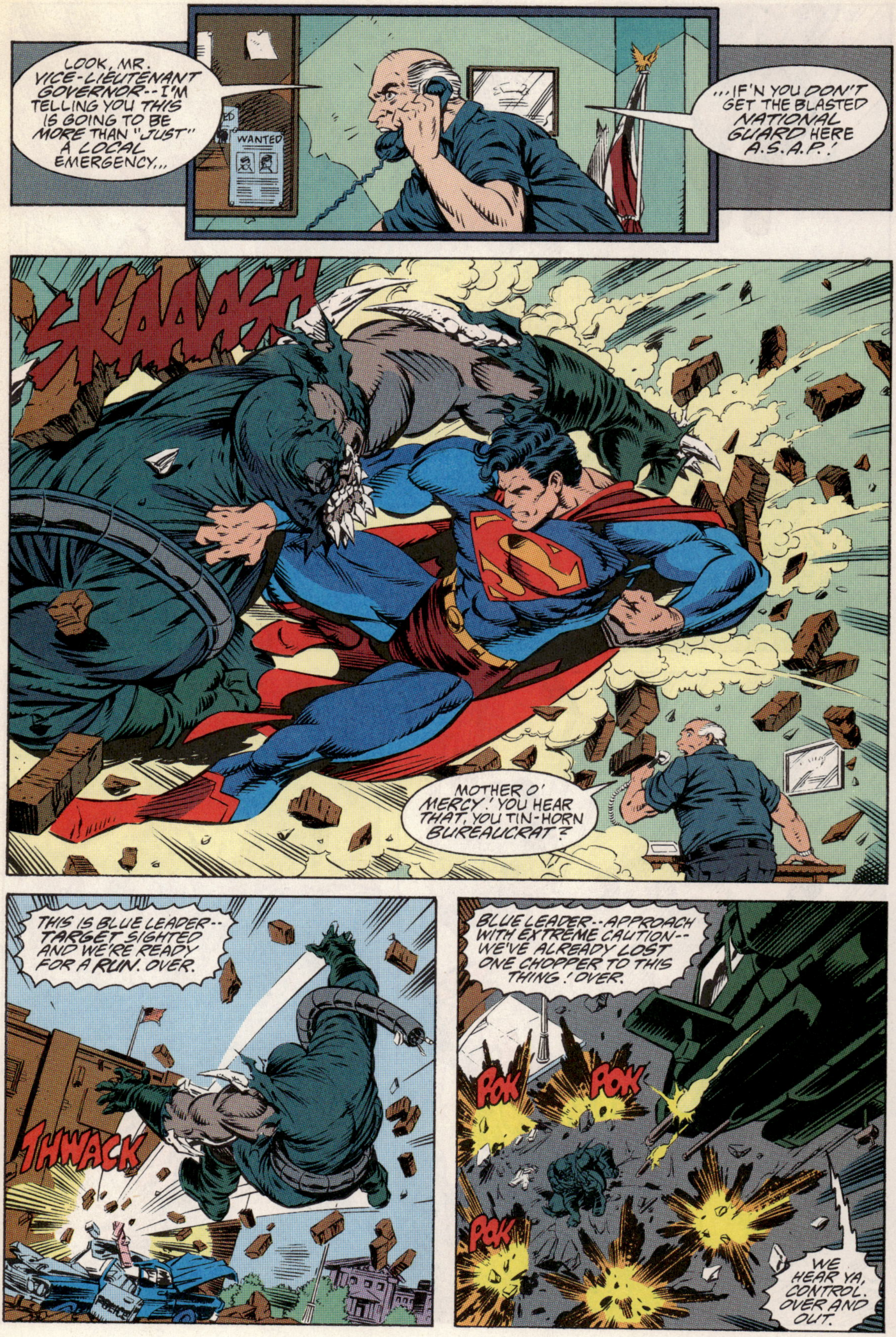
LOOK, MR. VICE-LIEUTENANT GOVERNOR--I'M TELLING YOU THIS IS GOING TO BE MORE THAN "JUST" A LOCAL EMERGENCY...
WANTED
...IF'N YOU DON'T GET THE BLASTED NATIONAL GUARD HERE A.S.A.P.!
SKAAASH
MOTHER O' MERCY! YOU HEAR THAT, YOU TIN-HORN BUREAUCRAT?
THIS IS BLUE LEADER--TARGET SIGHTED AND WE'RE READY FOR A RUN. OVER.
THWACK
BLUE LEADER--APPROACH WITH EXTREME CAUTION--WE'VE ALREADY LOST ONE CHOPPER TO THIS THING! OVER.
POK
POK
POK
WE HEAR YA, CONTROL. OVER AND OUT.

MAIN ST.
R-RRANK
THWOCK
THIS "BUBBA'S" TAKING EVERYTHING THE CHAINGUN'S RAINING ON HIM WITH A SMILE!
CHANK
WHAT THE--? WE'VE BEEN HIT!
NO... WE'VE BEEN SPEARED...
THE BACK-UP CONTROL SYSTEM'S NOT GONNA DO SQUAT WITH "BUBBA" DOWN THERE DIRECTING OUR FLIGHT!
HO-LEEEEE-- NO TIME TO BAIL! MAYDAY MAYDAY!

WHU? WHO'S THAT?
RELAX, SOLDIER. YOU AND YOUR CO-PILOT ARE GOING TO BE OKAY...
...THOUGH I DON'T THINK THIS TOWN'S CITY HALL WILL BE OPEN FOR BUSINESS ANY TIME SOON!
SHOOM!
TOWN HALL
"NOW EXCUSE ME, GUYS, BUT THERE ARE A DOZEN PEOPLE TRAPPED IN THAT WRECKAGE WHO NEED MY HELP!"
"STAND THERE AMID THE DESTRUCTION AND REVEL IN IT, WARRIOR!"
YOUR MOTIVES ARE UNCLEAR AS YET...
...BUT IF IT IS BATTLE YOU CRAVE, I, MAXIMA, AM PLEASED TO OBLIGE!
KA-POW
"YOU CAN'T JUST BARGE IN LIKE THAT, LADY!"

BUT IT'S AN EMERGENCY!
LOOK, LADY--THAT RED LIGHT OVER THE DOOR MEANS THEY'RE TAPING--THE CAMERAS ARE ROLLING.
GET IT? YOU CAN'T JUST BUST IN ON THEM!
CAN YOU AT LEAST TELL ME HOW I CAN GET A MESSAGE TO SOMEONE IN THERE?
STUDIO B
LOIS LANE? WHAT BRINGS YOU TO WGBS'S STUDIOS?
CAT GRANT! THANK GOD. A FAMILIAR FACE!
LOOK, JIMMY OLSEN'S IN THERE, AND HE'S NEEDED ON AN ASSIGNMENT.
STUDIO B
THEY'RE TAPING "THE TURTLE-BOY SHOW," MS. GRANT.
HE COULD LOSE HIS JOB!
I'LL TAKE RESPONSIBILITY IF WE DISRUPT ANYTHING, BUT JUST KEEP YOUR VOICE DOWN LOW, LOIS.
THIS HAS TO DO WITH SUPERMAN AND THE DESTRUCTION NORTH OF HERE, RIGHT?

HI, "TURTLE-BOY."
WHAT'S GOING ON?
JIMMY-- THE CHIEF WILL HAVE YOUR HIDE! YOUR LUNCH HOUR ISN'T THREE HOURS LONG!
LOIS!
ERR, TAPING RAN A LITTLE LONG, BUT THIS IS MY FIRST TV SHOW.
WHY ARE YOU WHISPERING?
PERRY WANTS US TO COVER THIS "DOOMSDAY" INCIDENT. NED'S WAITING FOR US AT THE HELI-PAD! HURRY UP!
YOU TWO BETTER LOOK AT SOMETHING FIRST. GBS IS ABOUT TO INTERRUPT "THE BRAVE AND THE BOLD" FOR A NEWSBREAK.
COME ON, "TURTLE-BOY!"
THAT'S MISTER "TURTLE-BOY" TO YOU.
SURE, FINE. JUST LEAVE BEFORE WE'RE FINISHED, WHY DON'T YOU?
HI, LEON. MIND IF WE WATCH?
NOPE.
THIS IS A GBS NEWSBREAK. I'M STEVE LOMBARD.
THE DESTRUCTIVE FORCE KNOWN AS "DOOMSDAY" HAS LEFT THIRTY DEAD IN ITS WAKE...
...HUNDREDS HAVE BEEN INJURED, INCLUDING MEMBERS OF THE FAMED JUSTICE LEAGUE.
DOOMSDAY
IT APPEARS "DOOMSDAY" IS ON A STRAIGHT PATH CROSSING FROM OHIO THROUGH NEW YORK STATE...

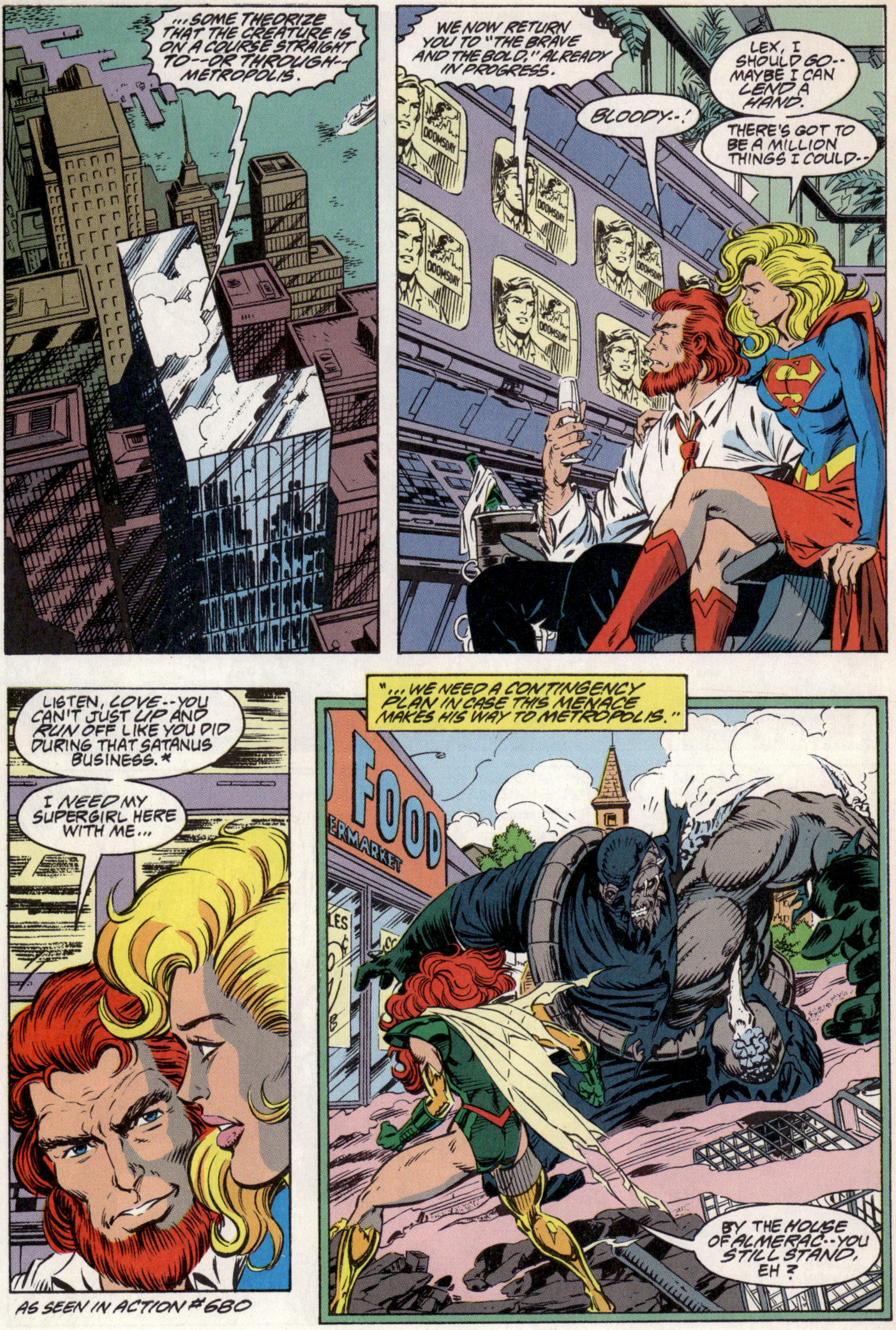
...SOME THEORIZE THAT THE CREATURE IS ON A COURSE STRAIGHT TO--OR THROUGH--METROPOLIS.
WE NOW RETURN YOU TO "THE BRAVE AND THE BOLD," ALREADY IN PROGRESS.
BLOODY--!
LEX, I SHOULD GO-- MAYBE I CAN LEND A HAND.
THERE'S GOT TO BE A MILLION THINGS I COULD--
DOOMSDAY
LISTEN, LOVE--YOU CAN'T JUST UP AND RUN OFF LIKE YOU DID DURING THAT SATANUS BUSINESS.*
I NEED MY SUPERGIRL HERE WITH ME...
*AS SEEN IN ACTION #680
"...WE NEED A CONTINGENCY PLAN IN CASE THIS MENACE MAKES HIS WAY TO METROPOLIS."
FOOD
ERMARKET
BY THE HOUSE OF ALMERAC--YOU STILL STAND, EH?

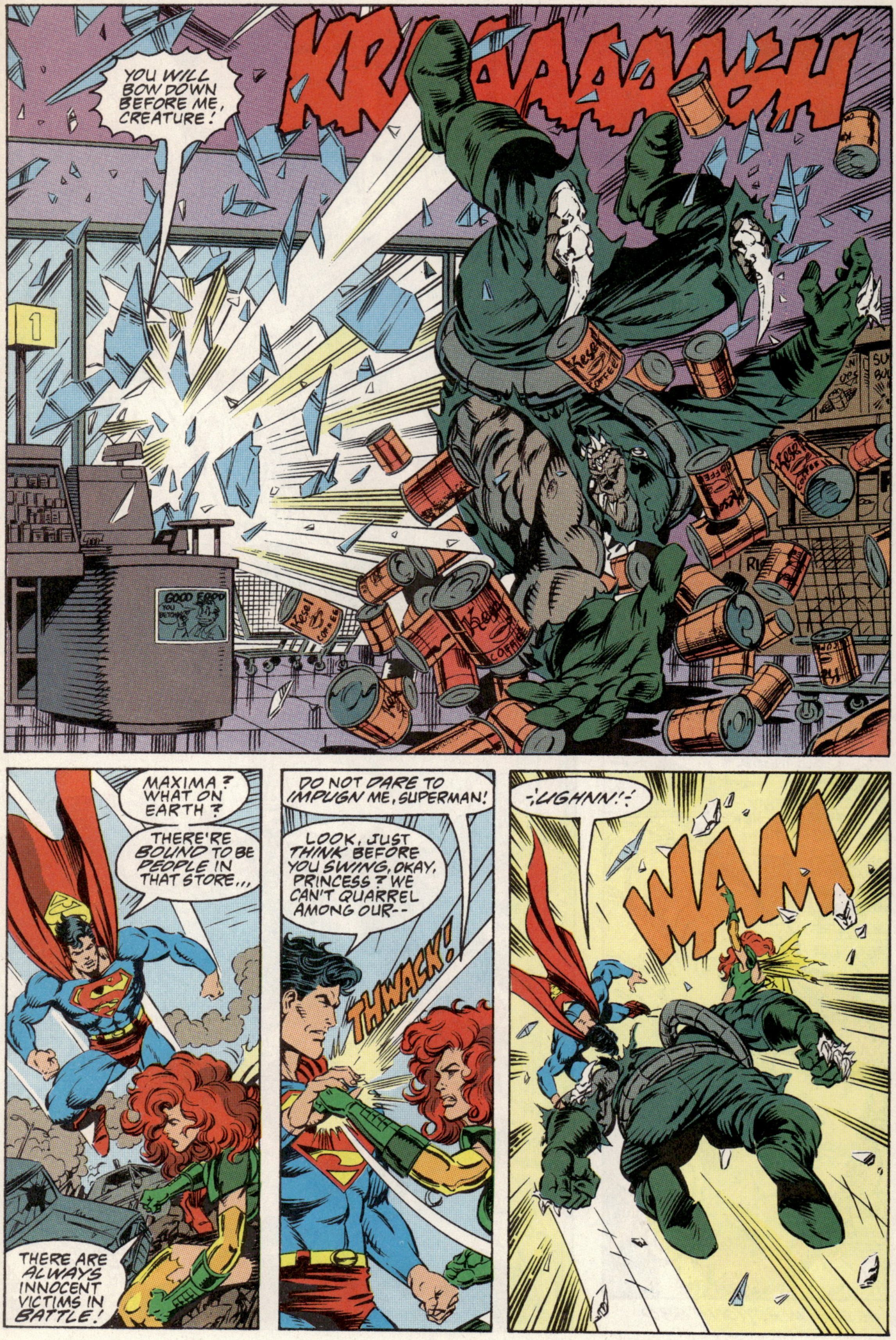
KRAAAAAASH
YOU WILL BOW DOWN BEFORE ME, CREATURE!
MAXIMA? WHAT ON EARTH?
THERE'RE BOUND TO BE PEOPLE IN THAT STORE...
THERE ARE ALWAYS INNOCENT VICTIMS IN BATTLE!
DO NOT DARE TO IMPUGN ME, SUPERMAN!
LOOK, JUST THINK BEFORE YOU SWING, OKAY, PRINCESS? WE CAN'T QUARREL AMONG OUR--
THWACK!
-UGHNN!-
WAM

WUF HURRY, MAXIMA--HIT HIM WITH EVERYTHING YOU'VE UGH GOT!
I UH CAN'T HOLD--
SMACK
HOW COULD HE MOVE SO FAST?
HE'LL NOT CATCH ME BY SURPRISE--
THWOCK
--AGAIN-- OOF!
EAG
EAGLE
SHAM

YOUR ONSLAUGHT DOES LITTLE BUT STIMULATE ME, CREATURE!
MAXIMA WELCOMES THIS, FOR ONLY WHEN A WARRIOR FACES DEATH CAN A CONFLICT BE DEEMED TRULY WORTHY!
MAXIE'S REVELING IN THIS, AND DOOMSDAY DOESN'T SEEM TO BE SLOWING DOWN MUCH...
SLAM
...BUT I DON'T KNOW HOW MUCH LONGER I CAN KEEP THIS UP!
FOR THE TIME BEING, I'D BETTER WORRY ABOUT ALL THIS GASOLINE GUSHING UP!
HOLD HIM TIGHT, KRYPTONIAN-- MAXIMA WILL NOT MISS AGAIN!
MAXIMA! THAT LIGHT POLE'S GOING TO SPARK!
ZZAKT

KAA-BOOOOOM
"I SAW THE THICK, BLACK SMOKE FROM A MILE AWAY.
"I DIDN'T NEED THE SOPHISTICATED TRACKING EQUIPMENT I HAD AT MY FINGERTIPS.
"I SIMPLY FOLLOWED THE SMOKE.

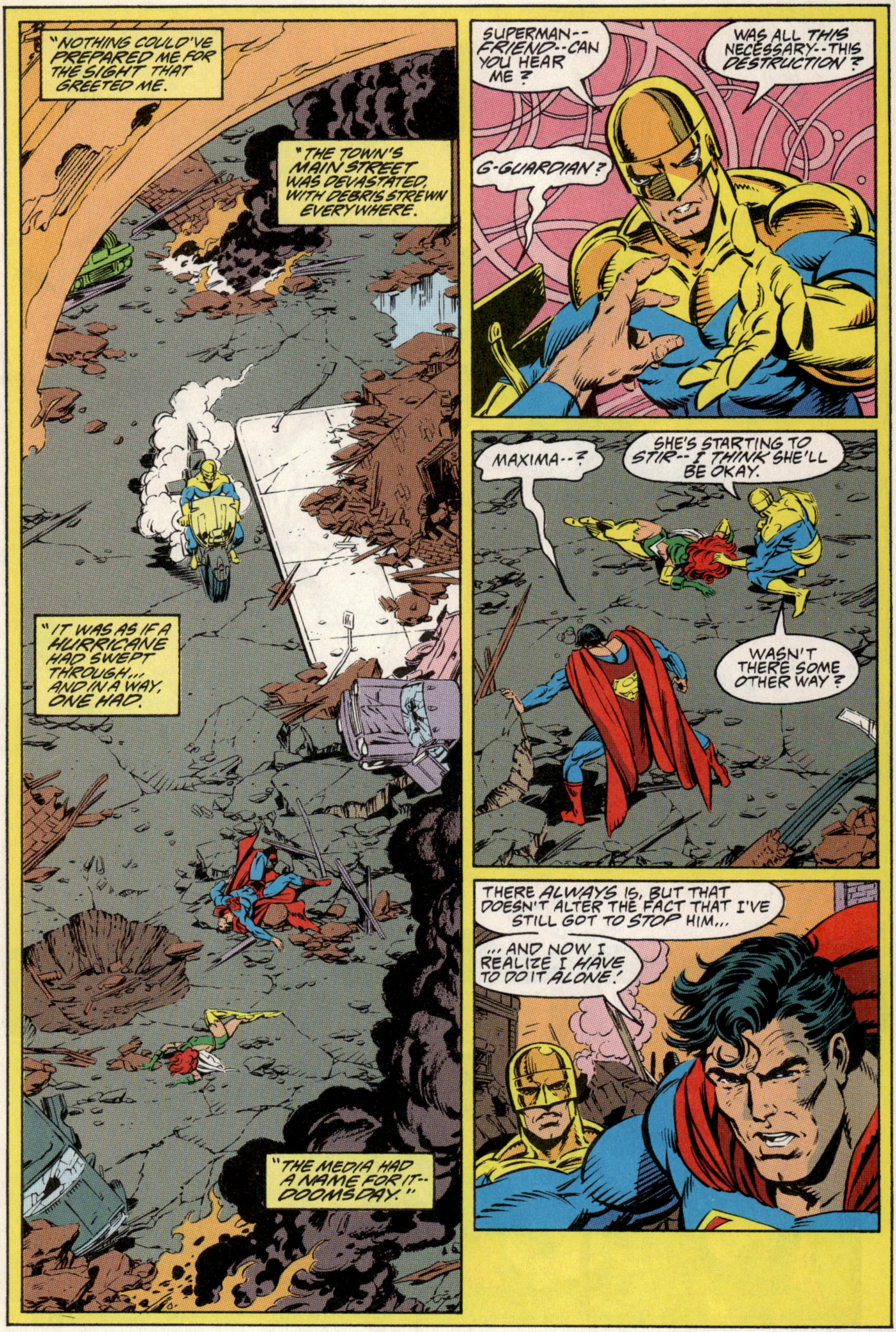
"NOTHING COULD'VE PREPARED ME FOR THE SIGHT THAT GREETED ME.
"THE TOWN'S MAIN STREET WAS DEVASTATED, WITH DEBRIS STREWN EVERYWHERE.
"IT WAS AS IF A HURRICANE HAD SWEPT THROUGH... AND IN A WAY, ONE HAD.
"THE MEDIA HAD A NAME FOR IT-- DOOMSDAY."
SUPERMAN-- FRIEND--CAN YOU HEAR ME?
WAS ALL THIS NECESSARY--THIS DESTRUCTION?
G-GUARDIAN?
MAXIMA--?
SHE'S STARTING TO STIR-- I THINK SHE'LL BE OKAY.
WASN'T THERE SOME OTHER WAY?
THERE ALWAYS IS, BUT THAT DOESN'T ALTER THE FACT THAT I'VE STILL GOT TO STOP HIM...
...AND NOW I REALIZE I HAVE TO DO IT ALONE!

...DOOMSDAY IS NEAR!
A BATTLE THAT HAS RAGED ACROSS HALF THE NATION SINCE MIDDAY, HAS LEFT SEVERAL MEMBERS OF JUSTICE LEAGUE AMERICA SERIOUSLY INJURED.
THE BLUE BEETLE IS REPORTED TO BE COMATOSE, AND BOOSTER GOLD SERIOUSLY INJURED FOLLOWING... ONE MOMENT!
SPECIAL REPORT
THIS JUST HANDED ME... THE VILLAGE OF GRIFFITH IN UPSTATE KIRBY COUNTY WAS ROCKED BY AN EXPLOSION MOMENTS AGO, AS SUPERMAN AND MAXIMA FOUGHT TO STOP THE CREATURE-- DUBBED DOOMSDAY--
CAMCORDER FOOTAGE
--WHOSE RAMPAGE HAS BROKEN THE JLA AND LEFT A TRAIL OF DEATH AND DESTRUCTION BEHIND HIM. DESPITE THEIR EFFORTS, HOWEVER, THE CREATURE IS REPORTEDLY STILL ON THE LOOSE.
CIVIL DEFENSE UNITS IN CITIES ALL ALONG THE EASTERN SEABOARD ARE ON ALERT, AS AUTHORITIES TRY TO DETERMINE IF...
DOOMSDAY... MUST STOP DOOMSDAY...
PLEASE, MAXIMA... TAKE IT SLOW AND EASY. YOU'VE SUFFERED A PRETTY SERIOUS CONCUSSION.
SORRY I DIDN'T GET HERE SOONER, SUPERMAN.
I DOUBT THAT YOU COULD HAVE HELPED US AVOID THIS, GUARDIAN.
WE'VE NEVER FACED ANYTHING QUITE LIKE THIS BEFORE.
ROGER STERN WRITER • JACKSON GUICE & DENIS RODIER ARTISTS
BILL OAKLEY LETTERER • GLENN WHITMORE COLORIST • JENNIFER FRANK ASSISTANT • MIKE CARLIN EDITOR
SUPERMAN created by JERRY SIEGEL & JOE SHUSTER

MAXIMA'S RIGHT... DOOMSDAY MUST BE *STOPPED*! HE'S A THREAT TO EVERY LIVING THING!
BUT SHE'S IN NO CONDITION TO DEAL WITH HIM.
GET HER TO A *HOSPITAL*, GUARDIAN.
I'LL STOP DOOMSDAY... IF IT'S THE LAST THING I DO!
LETHAL
MY GOD... LOOK AT THAT!
TRACKING DOOMSDAY IS LIKE FOLLOWING THE PATH OF A *TORNADO*... THERE'S UTTER DEVASTATION EVERYWHERE HE TOUCHES DOWN.
THERE ARE A HALF-DOZEN MAJOR URBAN CENTERS IN THIS REGION... OVER *25 MILLION* HUMAN LIVES ENDANGERED BY THAT MONSTER!
I WISH I KNEW WHERE DOOMSDAY CAME FROM...

"... I'VE NEVER SEEN ANYTHING --
KTOOM
"-- ON EARTH OR OFF IT -- TO EQUAL HIM FOR SHEER BRUTE STRENGTH!
"IT WAS PAINFUL TO SEE WHAT HE'D DONE TO GUY GARDNER.
FRAKAMM!
"IF ANYTHING, HE'S MORE SINGLE-MINDED THAN DRAAGA WAS... AND HE SEEMS EVEN MORE IRRATIONAL THAN LOBO, IF THAT'S POSSIBLE."
GRAUHRRR!
"THERE'S A FRIGHTENING THOUGHT. LOBO'S GIVEN ME A HARD TIME MORE THAN ONCE... BUT LOBO'S POWER, HIS RAGE, DOESN'T BEGIN TO COMPARE TO DOOMSDAY'S!

"THERE'S NO DISCERNIBLE PATTERN TO HIS MOVEMENTS--"
MY GOD, THE OVERPASS HAS COLLAPSED!
WHAT'S THAT COMING OUT--?
NO! CAN'T STOP IN TIME! I'M GONNA HIT--
HRAURR!
--HIM! HUH? HE...?
WHAT DID--? HOW DID HE--?
OMIGOD... I DON'T BELIEVE THIS!
I'M... FLYING?
"--HE JUST SEEMS TO WANDER FROM PLACE TO PLACE, ATTACKING WHATEVER CATCHES HIS EYE."
RAUR?

THIS CAN'T BE HAPPENING. I MUST BE DREAMING.
THAT'S IT... I'VE DOZED OFF AT THE WHEEL. GOT TO WAKE UP BEFORE I HAVE AN ACCIDENT!
WAKE UP, CHARLIE...
...WOW... MUST BE NEARLY A MILE UP. EVERYTHING LOOKS SO PRETTY FROM UP HERE... SO... REAL.
WHAT IS THE MATTER WITH ME?! WAKE UP, ALREADY!!
FEELS LIKE WE'RE SLOWING DOWN. OMIGOD, THE CAR'S TIPPING BACKWARDS!
THIS IS NO DREAM. I'M GONNA DIE.
IT'S OKAY! I'VE GOT YOU!
YOU'VE GOT ME. HEH SURE.
POOR GUY. HE MUST BE HYSTERICAL.
SIR? DON'T BE AFRAID-- EVERYTHING'S GOING TO BE ALL RIGHT. I'M SUPERMAN.
S-S-SUPER...MAN?
NO CHANCE OF THAT, SIR. KEEP TALKING... AND TAKE LONG, SLOW, DEEP BREATHS. DON'T GO INTO SHOCK ON ME NOW.
I HOPE YOU'RE REAL... OTHERWISE, I KNOW I'M DEAD!
I'VE BEEN SEARCHING FOR THE CREATURE WHICH MUST HAVE ATTACKED YOU. CAN YOU REMEMBER ANYTHING ABOUT HIM... ANYTHING AT ALL?

CREATURE? I... YEAH! HE... WAS BIG... CAME RIGHT AT ME. H-HE GRABBED HOLD OF MY CAR AND JUST... THREW IT!
IT HAPPENED SO FAST... DIDN'T SEEM REAL AT FIRST. WHAT... WHAT IS HE, SUPERMAN?
I WISH I KNEW. HE CAME FROM OUT OF NOWHERE -- DESTROYING THINGS AT RANDOM -- APPARENTLY FOR THE SHEER HELL OF IT!
THEN... YEAH... IT MUSTA BEEN HIM THAT COLLAPSED THE OVERPASS!
OVERPASS?!
"DAMN. I DON'T SEE ANY SURVIVORS AMONG THE WRECKAGE.
"LOOKS LIKE THERE'RE DOZENS OF CHAIN-REACTION FENDER-BENDERS UP AND DOWN BOTH HIGHWAYS... LOTS OF MINOR INJURIES THERE.
"AH-- THERE'S A STATE TROOPER ON THE SCENE. GOOD. AND I HEAR RESCUE VEHICLES ON THE WAY--!"
OH, NO!
WHAT'S WRONG? WHAT DO YOU SEE?
MORE TROUBLE... TERRIBLE TROUBLE.
I'M NEEDED--! I'LL SET YOU DOWN NEAR THAT STATE TROOPER.
TELL HER TO CALL FOR MORE RESCUE TEAMS! WE'LL NEED THEM--

"-- AT THE SHOPPING PLAZA ON THE NORTHWEST SIDE OF MIDVALE."
RUN! RUN FOR YOUR LIVES!
MY CAR--!
LEX-MART
MY KIDS!! OGOD, WHERE ARE MY KIDS?!
ATTENTION, LEX-MART SHOPPERS. THIS IS AN EMERGENCY SITUATION...
...PLEASE EXIT THE STORE IN A CALM AND ORDERLY FASHION.
GRAUR!
CHILD
H-E-L-P!
HEY, YOU!
URR?
VIDEO
YEAH, I'M TALKING TO YOU! COME CLOSER--

--YOU DON'T WANT TO MISS A SINGLE MOMENT OF THE GREATEST SPECTACLE IN THE HISTORY OF PROFESSIONAL WRESTLING!
WUNK THUD
I'M TALKIN' TAG TEAMS! I'M TALKIN' STEEL CAGES!
VIDEO CENTER
PANAVISION
I'M TALKIN' KNOCK-DOWN, DRAG-OUT GRUDGE MATCHES!
Lex-MART SPECIAL LIST PRICE
72 INCH BIG SCREEN GALAXY TV
$2,995.00
REMOTE CONTROLLED
UHN?
I'M TALKIN' 'BOUT WAR-BASH 9000! THIS WEEKEND! AT THE METROPOLIS ARENA!
I'M MAJOR MAYHEM, TROOPERS-- AND I'M OUT FOR BLOOD!
I'M TAKIN' ON THE MIGHTY GORILLA! UGLY BEN STUDLY! AND THE MASKED BONE-CRUSHER! AND--I--WILL --PREVAIL!
THIS TIME... IT'S WAR!!!
CAMCOR
THE METROPOLIS
ARENA
PRO-WRESTLING AS YOU'VE NEVER SEEN IT BEFORE! THIS WEEKEND AT METROPOLIS ARENA... METROPOLIS ARENA... METROPOLIS ARENA!!
NOW... WHERE ARE YA GONNA GO?
WAR-BASH 9
GET READY FOR BLOOD!!
MHH-TRR-PLSS?

DOOMSDAY!!
URRGH?
KTOOM!
HOLY--! WHAT WAS *THAT*?!
I... I THINK THE RED-AN'-BLUE BLUR IS SUPERMAN. I DON'T *WANNA* KNOW WHAT THE OTHER THING IS!
AMSCRAY, YOU TWO! WE GOTTA EVACUATE THIS LOADIN' DOCK--ON THE DOUBLE!

HAH-HA-HAH-HA-HA!
OW! THIS IS... INSANE!
I'D SWEAR... THE HARDER I FIGHT... THE MORE DOOMSDAY LIKES IT!
HE'S BEEN FIGHTING MOST OF THE DAY, BUT HE STILL SEEMS AS EAGER--AND AS STRONG--AS EVER!
IF HE HAS ENERGY RESERVES AS EXTENSIVE AS MINE, I MAY BE IN TROUBLE!
WUP-WUP-WUP!
EH? 'COPTERS...
...THE ONE IN THE LEAD IS THE DAILY PLANET'S FLYING NEWSROOM--
"--OH, LORD, AND LOIS AND JIMMY ARE ON BOARD! I HOPE THE PILOT KEEPS HIS DISTANCE!"
THAT'S DOOMSDAY? WOW, HE'S A BIG ONE!
VERY BIG. BE CAREFUL, CLARK.
"...THE MIDVALE LEX-MART STOOD IN RUINS AS SUPERMAN STRUGGLED WITH THE MYSTERIOUS CREATURE." END OF PARAGRAPH...

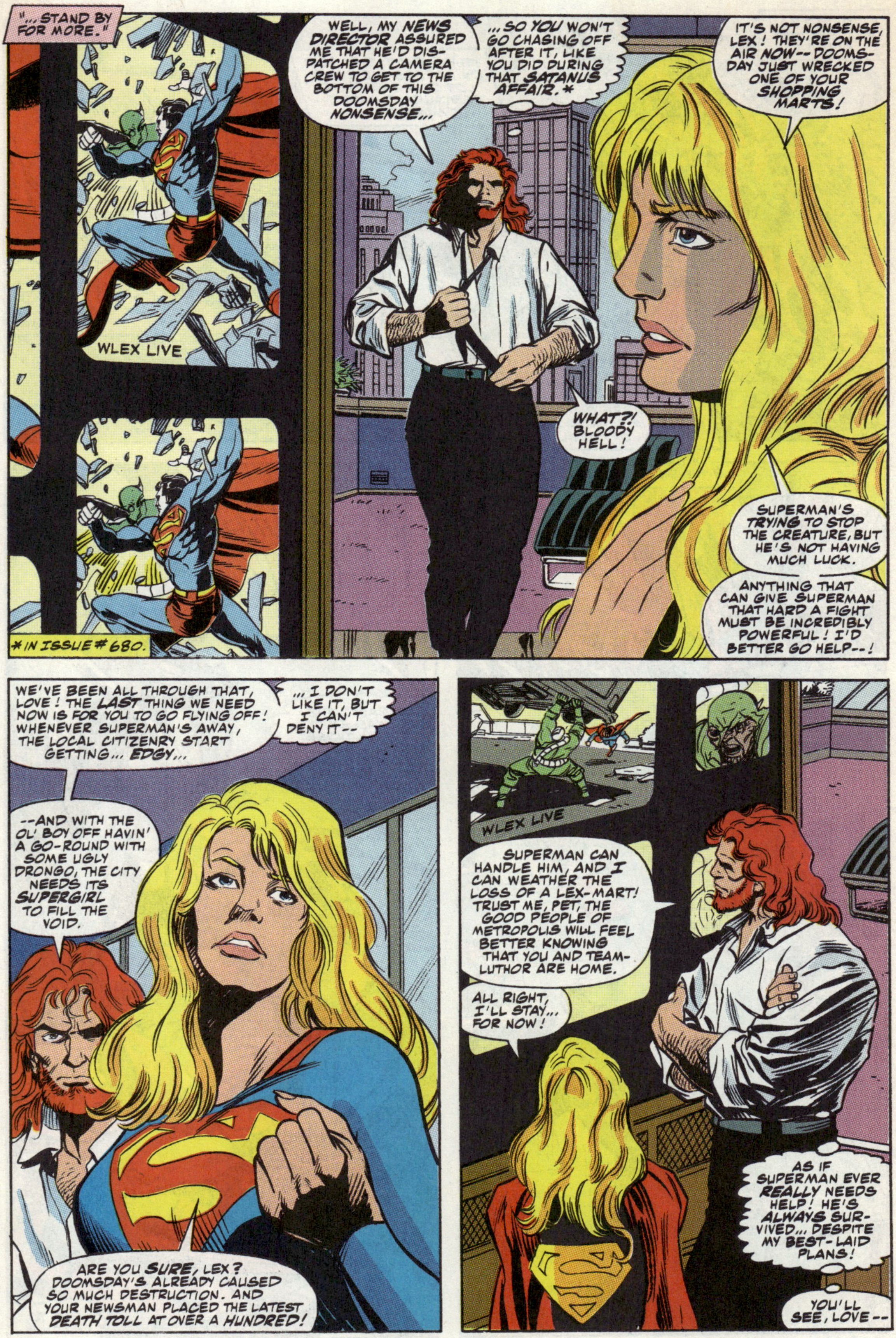
"...STAND BY FOR MORE."
WLEX LIVE
WELL, MY NEWS DIRECTOR ASSURED ME THAT HE'D DISPATCHED A CAMERA CREW TO GET TO THE BOTTOM OF THIS DOOMSDAY NONSENSE...
...SO YOU WON'T GO CHASING OFF AFTER IT, LIKE YOU DID DURING THAT SATANUS AFFAIR.*
IT'S NOT NONSENSE, LEX! THEY'RE ON THE AIR NOW--DOOMSDAY JUST WRECKED ONE OF YOUR SHOPPING MARTS!
WHAT?! BLOODY HELL!
SUPERMAN'S TRYING TO STOP THE CREATURE, BUT HE'S NOT HAVING MUCH LUCK.
ANYTHING THAT CAN GIVE SUPERMAN THAT HARD A FIGHT MUST BE INCREDIBLY POWERFUL! I'D BETTER GO HELP--!
*IN ISSUE #680.
WE'VE BEEN ALL THROUGH THAT, LOVE! THE LAST THING WE NEED NOW IS FOR YOU TO GO FLYING OFF! WHENEVER SUPERMAN'S AWAY, THE LOCAL CITIZENRY START GETTING... EDGY...
...I DON'T LIKE IT, BUT I CAN'T DENY IT--
--AND WITH THE OL' BOY OFF HAVIN' A GO-ROUND WITH SOME UGLY DRONGO, THE CITY NEEDS ITS SUPERGIRL TO FILL THE VOID.
ARE YOU SURE, LEX? DOOMSDAY'S ALREADY CAUSED SO MUCH DESTRUCTION. AND YOUR NEWSMAN PLACED THE LATEST DEATH TOLL AT OVER A HUNDRED!
WLEX LIVE
SUPERMAN CAN HANDLE HIM, AND I CAN WEATHER THE LOSS OF A LEX-MART! TRUST ME, PET, THE GOOD PEOPLE OF METROPOLIS WILL FEEL BETTER KNOWING THAT YOU AND TEAM-LUTHOR ARE HOME.
ALL RIGHT, I'LL STAY... FOR NOW!
AS IF SUPERMAN EVER REALLY NEEDS HELP! HE'S ALWAYS SURVIVED... DESPITE MY BEST-LAID PLANS!
YOU'LL SEE, LOVE--

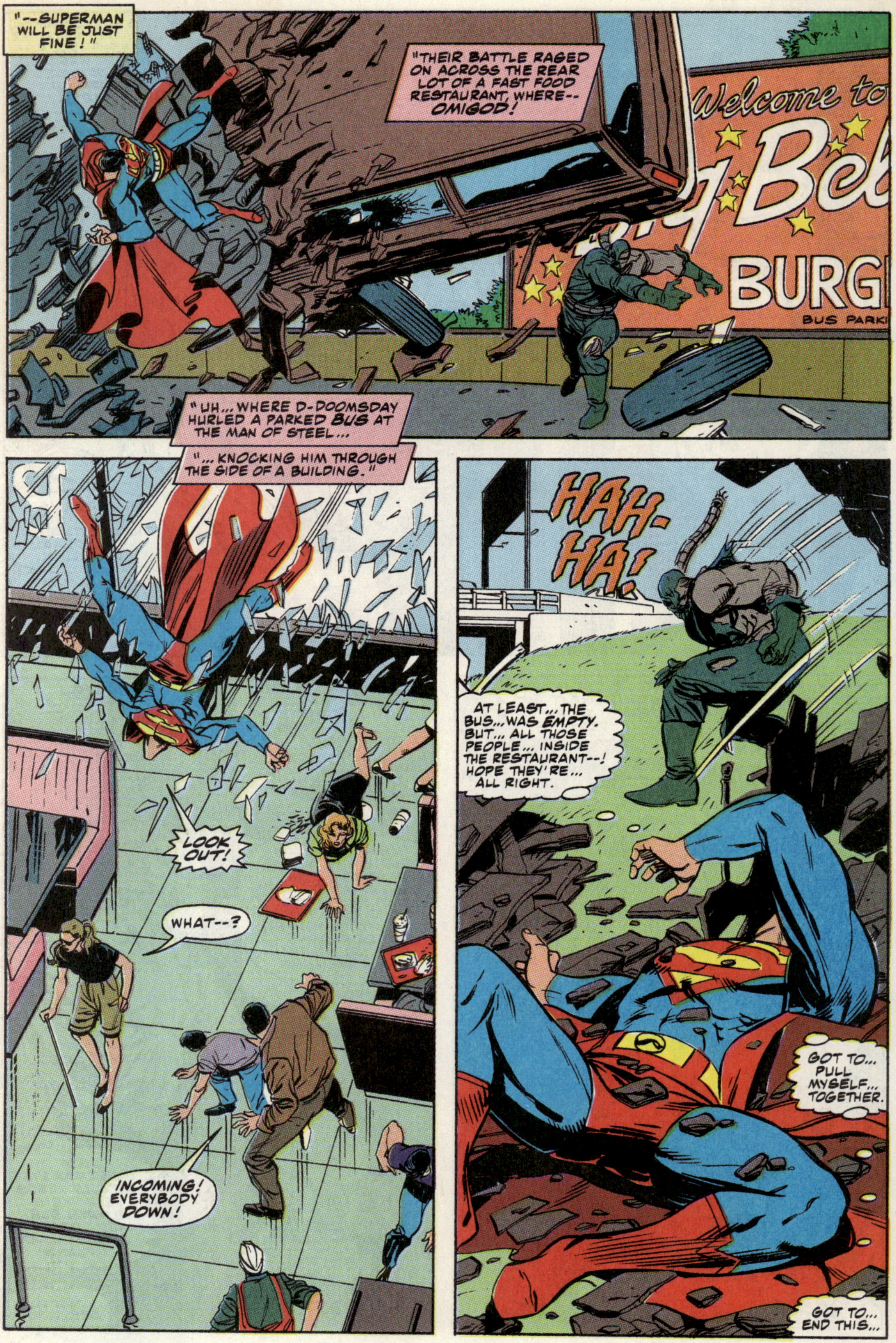
"--SUPERMAN WILL BE JUST FINE!"
"THEIR BATTLE RAGED ON ACROSS THE REAR LOT OF A FAST FOOD RESTAURANT, WHERE-- OMIGOD!
Welcome to
BURG
BUS PARK
"UH... WHERE D-DOOMSDAY HURLED A PARKED BUS AT THE MAN OF STEEL...
"...KNOCKING HIM THROUGH THE SIDE OF A BUILDING."
LOOK OUT!
WHAT--?
INCOMING! EVERYBODY DOWN!
HAH-HA!
AT LEAST... THE BUS... WAS EMPTY. BUT... ALL THOSE PEOPLE... INSIDE THE RESTAURANT--! HOPE THEY'RE... ALL RIGHT.
GOT TO... PULL MYSELF... TOGETHER.
GOT TO... END THIS...

HAH-HA...HUH? MHH-TRR-PLSS?
MHH-TRR-PLSS!! WHHR-BSSH!!
OH, NO.
METROPOLIS
60 MILES
NO!! HE REMEMBERS THAT STUPID COMMERCIAL! HE'S MAKING A CONNECTION--!
SIXTY MILES MIGHT AS WELL BE SIXTY PACES TO THIS MONSTER!
I CAN'T LET HIM GET ANY CLOSER! I CAN'T!!

GEEZ! SUPERMAN MUST'VE GOTTEN A SECOND WIND OR SOMETHING! I'VE NEVER SEEN HIM FIGHT SO HARD!
N-NEITHER HAVE I, JIMMY!
NEXT PARAGRAPH... "TAKING ADVANTAGE OF THE CREATURE'S MOMENTARY DISTRACTION--
"--SUPERMAN REDOUBLED HIS EFFORTS..."
GOT TO KEEP HIM OFF-BALANCE-- AND AVOID HIS REACH! HE MUST WEIGH CLOSE TO A TON...
...GOT TO USE THAT WEIGHT... BUILD UP ENOUGH MOMENTUM...
...TO HURL HIM AWAY FROM HERE... AWAY FROM METROPOLIS!

DOOMSDAY'S TAKEN EVERYTHING I'VE DISHED OUT SO FAR... MAYBE SLAMMING INTO THE HILLS AT A FEW HUNDRED MILES PER HOUR WILL SOFTEN HIM UP!
I HOPE SO! THIS IS ONE TIME I COULD REALLY USE SOME HELP... AND THE JLA IS INCAPACITATED!
I'M SURPRISED THAT YOUNG LEX LUTHOR HASN'T SENT SUPERGIRL TO HELP-- BY NOW HE SURELY KNOWS ABOUT DOOMSDAY SMASHING HIS STORE!
LEX THE SECOND IS A MORE ACTIVE TYPE THAN HIS FATHER WAS... I'D HALF-EXPECT HIM TO LEAD A TEAM-LUTHOR SQUADRON UP HERE HIMSELF!
OF COURSE, IF HIS FATHER WERE STILL ALIVE, I'D HAVE HALF-EXPECTED THE OLD MAN TO HAVE ENGINEERED THIS DOOMSDAY MONSTER!

I DON'T KNOW IF I CAN CATCH UP WITH THEM, MS. LANE, NOT AS FAST AS THEY'RE GOING!
JUST DO YOUR BEST, GARRET.
METROPOLIS ISN'T THAT FAR... I'LL BET SUPER-MAN'S TRYING TO KEEP DOOMSDAY AWAY FROM THE CITY.
WELL, HE'S HEADED IN THE RIGHT DIRECTION...

"... NOT MUCH TO WORRY ABOUT THERE. NO ONE'S ALLOWED MUCH UP INTO THOSE HILLS.
BUH-BOOM!
"EVEN A LOT OF THE AIR-SPACE IS RESTRICTED...

"... I THINK SOME SORT OF FEDERAL PRESERVE IS TUCKED AWAY UP THERE."
BUH-BOOM!
@#*%!!
WHAT'S GOING ON?! IS THIS AN EARTHQUAKE?!
CADMUS
PROJECT
INCONCEIVABLE! THIS IS ONE OF THE MOST GEOPHYSICALLY STABLE REGIONS ON THE CONTINENT! NO...
...THE PROJECT MUST BE UNDER SOME MANNER OF BOMBARDMENT!
TAKE IT EASY, WESTFIELD! WE'LL GET TO THE BOTTOM OF THIS.
YES... YES, YOU'RE RIGHT, JOHNSON... WE MUST!
THE GUARDIAN WOULD HAVE TO BE AWAY! THIS IS INORDINATELY INOPPORTUNE... UNLESS... YOU DON'T SUPPOSE--? NO, THE LEVEL OF COINCIDENCE IS FAR TOO GREAT...
"... AND YET, I CANNOT HELP BUT WONDER IF THIS SEISMIC DISRUPTION IS SOMEHOW RELATED TO THAT NEARBY MONSTER SCARE WHICH THE GUARDIAN IS INVESTIGATING."
UHHN... HHUNGH...

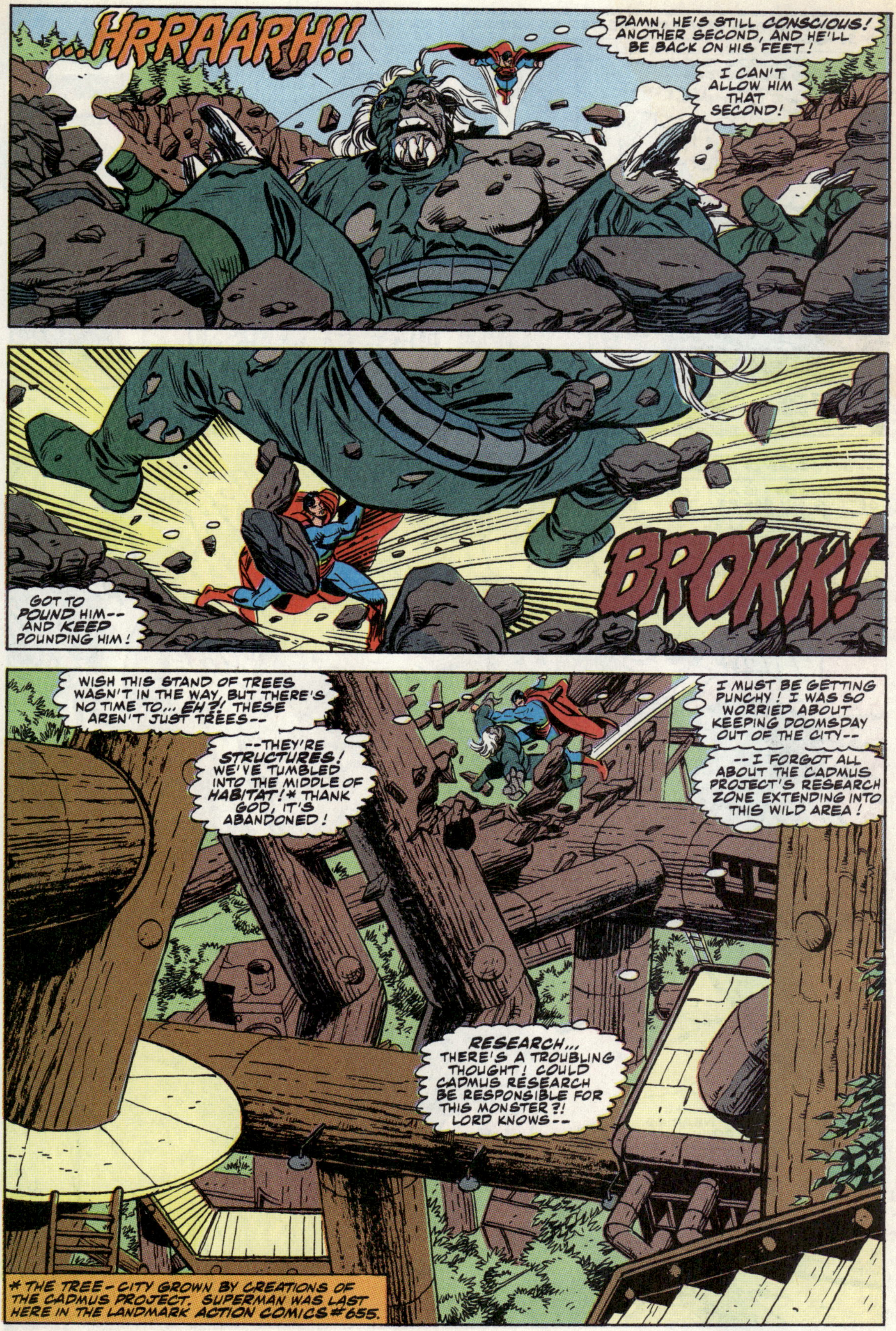
...HRRAARH!!
DAMN, HE'S STILL CONSCIOUS! ANOTHER SECOND, AND HE'LL BE BACK ON HIS FEET!
I CAN'T ALLOW HIM THAT SECOND!
GOT TO POUND HIM-- AND KEEP POUNDING HIM!
BROKK!
WISH THIS STAND OF TREES WASN'T IN THE WAY, BUT THERE'S NO TIME TO... EH?! THESE AREN'T JUST TREES--
--THEY'RE STRUCTURES! WE'VE TUMBLED INTO THE MIDDLE OF HABITAT!* THANK GOD, IT'S ABANDONED!
I MUST BE GETTING PUNCHY! I WAS SO WORRIED ABOUT KEEPING DOOMSDAY OUT OF THE CITY--
--I FORGOT ALL ABOUT THE CADMUS PROJECT'S RESEARCH ZONE EXTENDING INTO THIS WILD AREA!
RESEARCH... THERE'S A TROUBLING THOUGHT! COULD CADMUS RESEARCH BE RESPONSIBLE FOR THIS MONSTER?! LORD KNOWS--
* THE TREE-CITY GROWN BY CREATIONS OF THE CADMUS PROJECT. SUPERMAN WAS LAST HERE IN THE LANDMARK ACTION COMICS #655.

"--THEIR GENETICS LABS HAVE CREATED ALL MANNER OF BEINGS."
I THINK MAXIMA WILL BE OKAY... AS HARD A TIME AS SHE WAS GIVING THE DOCTORS--!
NEEP. NEEP. NEEP!
EH? THE ALERT SIGNAL? NOW WHAT?!
GUARDIAN--RETURN TO BASE AT ONCE!
WHAT IS IT, RODRIGUES? WHAT'S WRONG?
UNKNOWN. THE MOUNTAIN SEEMS TO BE UNDER ATTACK--
"--BY FORCES OF INCREDIBLE POWER!"
THIS :UNNGH: IS NOT :OW: GETTING ANY EASIER! JUST HITTING DOOMSDAY HURTS... AND HE DOESN'T SEEM... TO HAVE WEAKENED... ONE IOTA!
HRAH-HAH-HA!
THIS IS... JUST WEARING... ME DOWN. GOT TO... CHANGE MY TACTICS.
MAYBE IF I... HIT HIM WITH SOMETHING...

...Y-E-A-H... SOMETHING BIG!
HHUNH?
THESE SUPPORT COLUMNS... MAY BE ORGANIC... BUT THEY'RE HARD... AS TEMPERED STEEL!
WORTH A TRY--!
PTOOM!
GUARDIAN TO BASE! HABITAT... MY GOD, HABITAT IS COLLAPSING --LIKE A HOUSE OF CARDS!
AND SUPERMAN AND THAT MONSTER ARE SMACK IN THE MIDDLE OF IT!
I-IT'S BAD... I'M GOING IN FOR A CLOSER LOOK! I'LL KEEP THIS CHANNEL OPEN...

G-GUARDIAN? WHERE'D... YOU COME FROM? WHERE'S... DOOMSDAY?
BURIED... UNDER WHAT'S LEFT OF HABITAT. YOU BARELY GOT OUT OF IT YOURSELF.
YOU TOOK SOME TERRIBLE HITS IN THE COLLAPSE --WHY DIDN'T YOU FLY OUT OF IT?
TOO... WASTED. NEED TO REST... SOON AS I... MAKE SURE... DOOMSDAY'S... STOPPED.
RELAX, YOU GOT HIM THAT TIME.
HOPE SO... BUT HAVE TO BE CERTAIN. HARD TO SEE... THROUGH THE RUINS... EYES WON'T FOCUS. I... OH, NO.
KRAKATOOM!
GRAUHRARR!

HRRARH?!
HNNHH!
MHH-TRR-PLSS!
SKRAWK GUARDIAN? WE WERE CUT OFF FOR A MOMENT... CAN YOU READ ME?
WE DETECTED ANOTHER SHOCK... NO, TWO MORE SHOCKS! WHAT'S GOING ON THERE?
GUARDIAN? GUARDIAN!!
DR. RODRIGUES, THIS IS SECURITY! SELECT-SCAN RADAR HAS JUST CONFIRMED THE APPARENT LAUNCH OF AN OBJECT-- SOMEWHAT LARGER THAN MAN-SIZED--FROM THE WILD AREA...

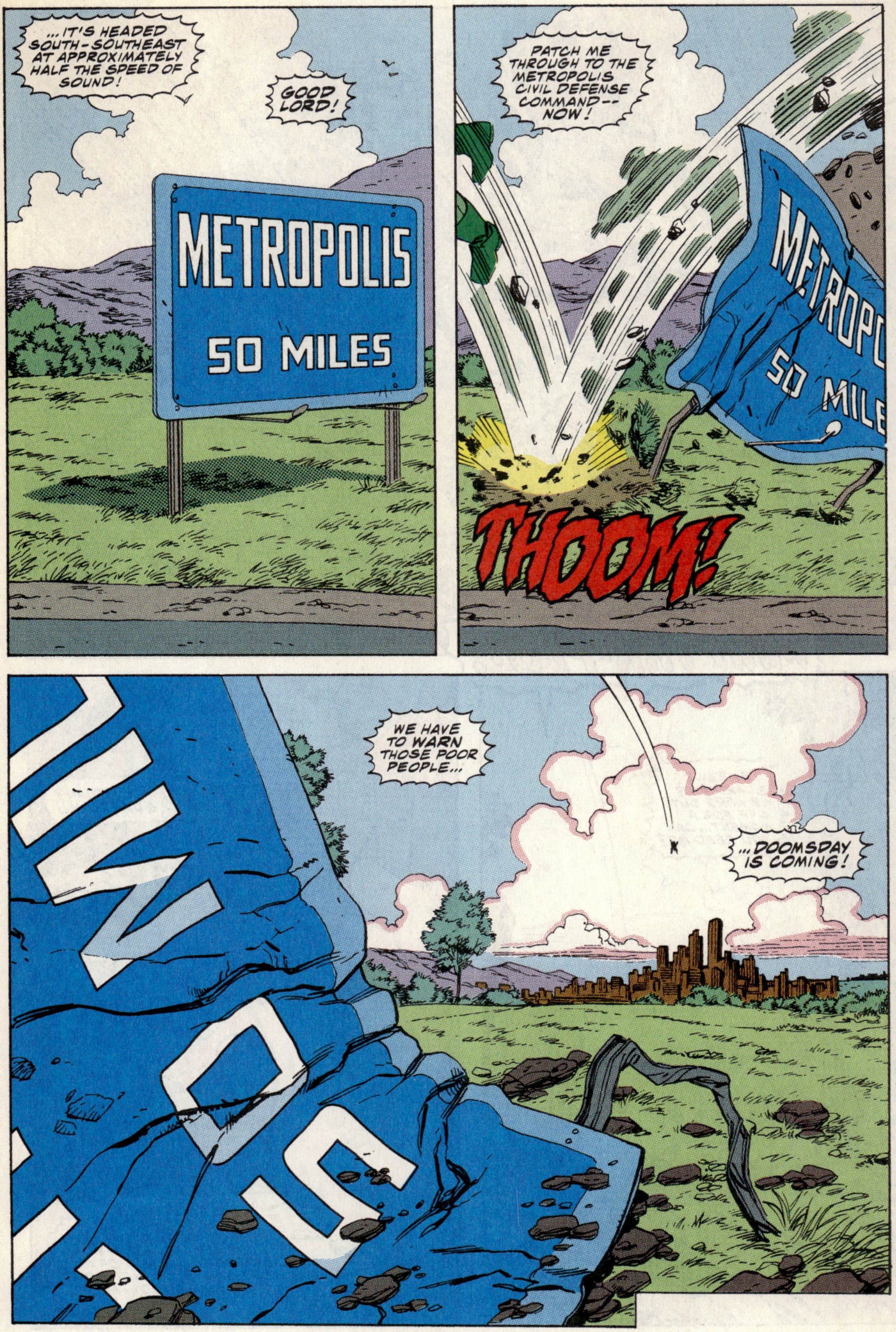
...IT'S HEADED SOUTH-SOUTHEAST AT APPROXIMATELY HALF THE SPEED OF SOUND!
GOOD LORD!
METROPOLIS
50 MILES
PATCH ME THROUGH TO THE METROPOLIS CIVIL DEFENSE COMMAND-- NOW!
THOOM!
WE HAVE TO WARN THOSE POOR PEOPLE...
...DOOMSDAY IS COMING!

"THE MONSTER CALLED DOOMSDAY ABANDONED THE FURROW OF DESTRUCTION HE PLOWED THROUGH OHIO AND WESTERN NEW YORK STATE...
"'...AND HEADED, IN TEN-MILE LEAPS, TOWARD THE EAST COAST AND METROPOLIS!' END OF PARAGRAPH."
AMERICAN
WHAT IS IT?
I DUNNO! JUST RUN!

MHH-TRR-PLSS!
METROPOLIS! IT SAID MET--
KRAKK
"METROPOLIS?! UH... LOIS... YOU SIGHTED HIM YET?"

KLANNG!
"NEGATIVE, FRAN. THE PLANET'S CHOPPER IS CHASING DOOMSDAY AS FAST AS IT CAN...
"...CAT GRANT'S WGBS CHOPPER IS ACCOMPANYING US...
HRARRH!
"...BUT SUPERMAN IS ON THE CREATURE'S TAIL.

"KEEP YOUR HEAD DOWN, FRAN. IF OUR CALCULATIONS ARE RIGHT, THAT MONSTER MUST BE REACHING METROPOLIS ABOUT NOW!"

DOOMSDAY IS HERE!
FORGET IT, DOOMSDAY! YOUR RAMPAGE STOPS HERE!
STORY: LOUISE SIMONSON
PENCILLER: JON BOGDANOVE
INKER: DENNIS JANKE
LETTERER: BILL OAKLEY
COLORIST: GLENN WHITMORE
ASSISTANT EDITOR: JENNIFER FRANK
EDITOR: MIKE CARLIN
SUPERMAN created by JERRY SIEGEL and JOE SHUSTER

HURH?!
AN' IF HE KNOWS WHAT'S GOOD FOR HIM, HE'LL STAY THERE.
I... HOPE SO, JOE. DOOMSDAY IS PROBABLY THE DEADLIEST FOE SUPERMAN HAS EVER FACED.
MAN, THIS MUST BE MY LUCKY YEAR. THOSE WERE SOME OF THE BEST SHOTS I'VE EVER GOTTEN.
KLIK-WRRR KLIK-WRRR
WE'VE GOT HIM, FRAN! NEW PARAGRAPH:
"DOOMSDAY'S RAMPAGE IN PARK RIDGE WAS CUT SHORT WHEN SUPERMAN GRABBED THE MONSTER...
"... ROCKETING HIM AWAY FROM METROPOLIS, TOWARD THE VACUUM OF SPACE." END OF PARAGRAPH.
DON'T SWEAT IT, LOIS. HE'S SUPERMAN, RIGHT--

"--HE'S GOTTA BE OKAY!"
GUARDIAN, ARE YOU ALL RIGHT?
DUBBILEX! WHAT... HAPPENED?
DOOMSDAY SMASHED HABITAT! YOU WERE FELLED BY THE RUBBLE.
AND SUPERMAN...?
EVEN NOW HE BATTLES THE CREATURE.
I'M AFRAID DOOMSDAY IS TOO BIG FOR SUPERMAN TO HANDLE ALONE.
DOOMSDAY MAY BE ONE OF OURS, GUARDIAN, A D.N.ALIEN... A CADMUS-DABNEY DONOVAN CREATION.
TRY TO MIND-READ THE CREATURE, DUB. FIND OUT. I JUST PRAY THAT CADMUS ISN'T RESPONSIBLE--
"--FOR THE DAMAGE THAT MONSTER'S CAUSED!"
WHACK
CREATURE'S AS AGILE... AS IT IS STRONG! TWISTED AWAY ...COULDN'T HOLD HIM...
KICKED ME... CAN'T BREATHE...
THERE IS NOTHING IN HIS MIND BUT ANGER--

NO THOUGHT BUT DESTRUCTION.
THERE IS NO WAY TO TELL WHERE HE CAME FROM. NOT THAT IT MATTERS.
WE'LL HAVE TO WORK TO STOP HIM IN ANY CASE. IF ANYONE CAN STOP HIM.
LOIS, LOOK! DOOMSDAY'S FREE!
I SEE HIM, JIMMY! OH, LORD, WHERE'S SUPERMAN?!
THERE... I HOPE HE'S ALL RIGHT. NEW PARAGRAPH, FRAN.
"FAR ABOVE METROPOLIS, THE MONSTER BROKE FREE, HURLING THE MAN OF STEEL MORE THAN A MILE TO THE GROUND..."
SUPERMAN'S DOWN! GET CLOSER, BLAST YOU! WGBS NEEDS A SHOT OF THIS!
KRAZH
THROMB

THAT'S OUR SON, JONATHAN! HE'S BEING BEATEN TO A PULP--
--AND THOSE TV REPORTERS ARE TREATING IT LIKE ENTERTAINMENT!
CLARK IS OUR BOY, MA. BUT TO THE WORLD HE'S SUPERMAN.
IT'S NOT THAT THEY'RE CALLOUS. 'LEAST THEY DON'T MEAN TA BE. IT'S JUST--
"--THEY DON'T THINK ANYTHING BAD CAN REALLY HAPPEN TO HIM!"
BETTER FINISH THIS QUICK... IF I'M GONNA LIVE TO FINISH IT AT ALL.
GOT AWAY AGAIN. HOW TO FIND HIM? SHOULDN'T BE HARD--

"--TO FOLLOW HIS PATH OF DESTRUCTION."
KRACK
THOMB
THE CEILING!
WATCH OUT!
WHAT CRASH THROUGH TO UNDERWORLD?!
WHAT THAT BE?
BIG SUCKER!
MAYBE COMIN' TA RESCUE THE PRISONERS?
THAT'S WHAT YOU FOOLS GET FOR DARING TO PUT WARWORLDERS ON TRIAL!!
:hurf!: ‹SMELL--›
‹--GAS?›
CRAKSH
TK-TK! ‹WHO CARE WHAT HE SMELL LIKE, KR'LI?! PLUG UGLY'S OUR TICKET TO FREEDOM!› TK-TK!

WE THANK YOU, SIR, FOR YOUR EFFORTS ON OUR BEHALF.
THOUGH, OF COURSE, YOU SERVE YOUR OWN ENDS AS WELL AS OURS.
AFTER ALL, IT IS WE WHO ARE THE CHAMPIONS OF THOSE LIKE YOU--
--THE POOR AND DOWN-TRODDEN--
ZPLATT!
ZZPLATT!
RHAR!
--REJECTS OF THE OVER-WOR--!
ZZPLATT!

IT WAS AS EASY TO FIND HIM AS I THOUGHT IT WOULD BE... UNFORTUNATELY.
ALL I HAD TO DO WAS LOOK FOR THE RUBBLE--
WHAT'S THAT--?!
SMELLS LIKE A GAS MAIN'S BEEN RUPTURED HERE!
ALL YOU UNDERWORLDERS --GET BACK-- FAST AND FAR AS YOU CAN!
DOOMSDAY LEAPS BUT DOESN'T FLY.
'LONG AS I HAVE HIM OFF THE GROUND, I HAVE SOME ADVANTAGE.
METROPOLIS POWER & LIGHT
CONDUIT #3164
STOP SQUIRMING, DAMN YOU!

THIS TIME YOU'RE NOT KICKING FREE OF ME!
SPARK
BOK!
RSSHHPP
METROPOLIS POWER&LIGHT
CONDUIT IN SIGHT
BA-DOOM!

AIIEEE!

AN EXPLOSION! OR A BOMB! IT... MUST HAVE LEVELED ALL OF NEWTOWN!
WHAT IS THAT CREATURE? WHY IS IT DESTROYING METROPOLIS?
PULL UP! PULL UP! WE'RE GOING TO CRASH!
I'M DOIN' WHAT I CAN! IT'S LIKE RIDIN' A FREAKIN' TIDAL WAVE!

YOU CAN TELL YOUR VIEWERS, MISS ANDERSON, THAT LEX LUTHOR DOESN'T KNOW WHAT DOOMSDAY IS OR WHERE HE CAME FROM--
--BUT IT HAS BECOME INCREASINGLY OBVIOUS WHY HE IS HERE!
THE CREATURE HAS A GRUDGE AGAINST SUPERMAN!
OH, LEX, DO YOU REALLY THINK SO?
IT SEEMS PROBABLE, DOESN'T IT, LOVE?
I'M LOATH TO ASK, BUT DOES METROPOLIS NEED A CHAMPION...
...WHO DRAWS SUCH... NEGATIVE ATTENTION?
DOES SUPERMAN'S PRESENCE HERE CAUSE MORE HARM THAN GOOD?
I-I THINK DOOMSDAY MAY BE MORE THAN SUPERMAN CAN HANDLE ALONE!
DON'T BE ANNOYED, LEX, BUT I HAVE TO HELP HIM!
VERY GENEROUS, LOVE. INDEED, I AGREE WITH YOU, METROPOLIS MUST BE PRESERVED!

AGGNHHH!
CAN'T--
SSLICKKT
--HOLD ON!
SUPERMAN!
DON'T WORRY, SUPERMAN! I HAVE HIM!

MY LORD IN HEAVEN! HE'S THROWN OFF SUPERMAN!
WHAT-- WHAT IS THAT CREATURE?
I SUSPECT IT'S A DOOMSDAY WEAPON, MILDRED...
...LEFT BEHIND BY WARWORLD TO DECIMATE THE EARTH IN CASE THEIR SWARM FAILED!
BLASH
WE FINALLY GOT DIS LASER CANNON SHOVED UP ON DA ROOF O' YOUR LAB, PERFESSER HAM...
...SO LET'S USE IT!!
AS SOON AS SUPERGIRL GETS OUT OF THE WAY, BIBBO!
YOU AND MILDRED KEEP YOUR FORCE BELTS BUCKLED TIGHT!
WHEN THE CREATURE FEELS THE BLAST, HE'S GOING TO BE ANGRY!
NEVER SEEN A CREATURE THIS POWERFUL. MUST BE SOME--

HRARR!
--KIND OF... ALIEN--!
WHRAMP
SHRAKKT
SUPERGIRL'S DOWN, PERFESSER! HIT 'IM NOW!
WE DID IT! WE GOT HIM!

BA-DOOM
HE'S FALLING, BUT--
EMIL, HE'S COMING STRAIGHT AT US!
IF HE'S TRYIN' TA FALL ON TOP O' US, HE'S GONNA MAKE IT! WE GOTTA--
--JUMP FOR IT!
HEY, PERFESSER, YOU SURE DESE FORCE BELTS GONNA PERTECT US?
IT'S A BIT LATE TO WORRY ABOUT THAT NOW, MR. BIBBOWSKI!
WHY WOULD HE DO THIS, EMIL? CAUSE DESTRUCTION FOR THE SHEER JOY OF IT?
EVERYTHING ABOUT THIS MONSTER IS CLEARLY BEYOND OUR UNDERSTANDING, MILDRED.

ANOTHER GAS MAIN... MUST HAVE... BEEN RUPTURED...
...SETTING OFF... SERIES OF EXPLOSIONS...
...ALL AROUND ME ...IN HOBS BAY!
BR-
BROMB!
LEGS FEEL LIKE JELLY!
HELP! SUPERMAN-- HELP!
IT'S KEITH--
--AND MYRA AT THE ORPHANAGE!
NOT GOOD FOR MUCH... BUT STILL... MAKE A PRETTY GOOD... SHIELD!
SUPERMAN! I KNEW HE'D COME!
BUT, MYRA, WHAT IS THAT DOOMSDAY MONSTER?
DID SOMEBODY BUILD HIM... LIKE A GIANT FRANKEN-STEIN?
I DON'T KNOW, BABY! FROM THE WAY HE'S BEHAVIN', I'D SAY HE'S THE DEVIL INCARNATE...
...USHERIN' IN THE END OF THE WORLD!
BRA-
DOOM!

HERE HE COMES, MAGGIE! UGLY SUCKER!
YOU'RE A MASTER OF UNDERSTATEMENT, TURP!
C'MON! NAIL HIM--
--NOW!
BLAM
BAM
BKAM
BRAM
DRAM
DBAM
KRAK
HE'S NOT STOPPING!
KDAM

YEOWW!
TURP!
MPD
SCU
SCU

GET MAGGIE AND THE S.C. UNIT OUT OF THE WAY, TURPIN! ON THE DOUBLE!
WHUUMPH!
SUPERMAN!
KSMAZH!
NOT SURE I CAN MAKE A DENT IN DOOMSDAY'S HIDE.
HE'S ALREADY MADE SEVERAL--

HRRARRRRGH!
--IN MINE!
HIT HIM WITH EVERYTHING I'VE GOT!
HOPE THAT... DESPITE HIS APPARENT TOUGHNESS... HE'S FEELING AS CHIPPED AROUND THE EDGES AS I AM!
KRAK!
SLIKT!
VISUAL SIGHTING CONFIRMED, SIR! DOOMSDAY IS RIGHT BELOW US!
OPEN FIRE!
CADMUS'S WEAPONS EXPERTS SWEAR THESE SHOCK CANNONS'LL DO THE TRICK! LET'S HOPE THEY'RE RIGHT!
THEY KEEP MOVING! WHAT IF THEY "DO THE TRICK" ON SUPERMAN?
THAT'S A CHANCE WE'LL HAVE TO TAKE, SOLDIER!
SEEMS LIKE EVEN SUPERMAN CAN'T STOP THAT THING--

--ALONE!
KRAKK!
YOUR RAMPAGE ENDS HERE, DOOMSDAY!
EVEN IF IT KILLS ME--
THOOM!

--METROPOLIS IS WHERE I HOLD THE LINE!

THE BATTLE HAS DEVASTATED THE BETTER PART OF AMERICA... LEAVING A PATH OF DESTRUCTION ALMOST AS LONG.
EARTH'S MIGHTIEST HEROES HAVE ALREADY FALLEN UNDER DOOMSDAY'S MURDEROUS BLITZKRIEG.
ONLY ONE HOPE... ONE MAN REMAINS.
IT STOPS HERE, MISTER! THIS INSANITY ENDS IN METROPOLIS!

FOCUS YOUR FIRE ON THE CREATURE!
THE WEAPONS BOYS AT PROJECT CADMUS SAY THESE SHOCK CANNONS CAN TAKE OUT A *TANK*!
HOW CAN WE HANDLE THIS GUY WHEN SUPERMAN CAN'T? THIS REALLY MIGHT BE--
MOVE IN CLOSER! WE'RE BROADCASTING THIS *LIVE*!
LEXPARK 4.50
ALL DAY SATURDAY
SUNDAY 5.00
ONE WAY
DAILY PLANET
HURFF!
SKRAKK

DOOMSDAY!

DAN JURGENS / BRETT BREEDING
Words & Pictures / Finished Art

JOHN COSTANZA / GLENN WHITMORE
Letters / Colors

JENNIFER FRANK / MIKE CARLIN
Assistant / Editor

SUPERMAN created by
JERRY SIEGEL & JOE SHUSTER

LOOK OUT!
TOO LATE, MS. LANE! OUR HYDRAULIC CABLES HAVE JUST BEEN SHEARED IN HALF!
BLAST! ALL THESE ONLOOKERS AND CHOPPERS ARE COMPLICATING MATTERS BY GETTING IN THE WAY!
IT'S GETTING NASTY HERE, MS. GRANT! WE BETTER BACK OFF!
NO WAY! WE ARE *NOT* GOING TO MISS THE STORY OF THE CENTURY!
Seth's

MAN, I DON'T BELIEVE THIS! I MAY BE GETTING THE GREATEST PICTURES EVER--
--BUT IT'S COSTING US HALF THE CITY!
SUPERMAN, ARE YOU ALL RIGHT? YOU LOOK SO...SO...
I'D LIKE TO GET YOU TWO AS FAR FROM DANGER AS POSSIBLE BUT I JUST DON'T HAVE THE TIME!
NO TELLING HOW MANY LIVES DOOMSDAY COULD TAKE WHILE I'M GONE!

Welcome to Metropolis
HE'S SHRUGGING OFF THAT CANNON FIRE LIKE IT WAS NOTHING! HE'S UNSTOPPABLE!
KRETCH!
PLEASE, MAYBE YOU SHOULD RETREAT AND GET HELP! IF JIMMY IS RIGHT...
RRARRH!
TOO LATE, LOIS. THE JLA HAS ALREADY FALLEN AND THERE ARE TOO MANY INNOCENTS IN JEOPARDY RIGHT NOW!
IT'S UP TO ME.

CLARK... I...
JUST REMEMBER... NO MATTER WHAT HAPPENS... I'LL ALWAYS LOVE YOU.
ALWAYS.

BUT NOW--
--DOOMSDAY GETS HIS.
WAIT! COME BACK. PLEASE COME BACK.
WOW, I DON'T THINK I'VE EVER SEEN THE BIG GUY SO FIRED UP!

NOBODY TEARS APART MY CITY AND GETS AWAY WITH IT!
I DON'T KNOW WHAT HOLE YOU CRAWLED OUT OF OR WHERE YOU CAME FROM--
--BUT I'M SENDING YOU BACK!
BRAKOMMM!

CAN YOU BELIEVE THAT? IF THIS KEEPS UP WE WON'T HAVE A CITY LEFT!
STAY CLOSE SO WE CAN KEEP UP THE BROAD-CAST!
THE WHOLE COUNTRY WILL WANT TO SEE SUPERMAN KICK THIS CREEP'S BUTT!
I WANT TO KEEP FIRING BUT IT'S IMPOSSIBLE! THEY'RE MOVING TOO FAST!

BAH HAA!
CAN'T--
UHF!
BRAKKT

STAY BACK, MISS LANE! WE DON'T NEED TO GET THAT CLOSE TO GET A DECENT PICTURE!
WE CAN'T WORRY ABOUT PICTURES! SUPERMAN IS IN TROUBLE--
--AND I INTEND TO HELP HIM!
DAILY PLANET
HURM?
ONE WAY

GRRRRRRR
I DON'T THINK WE'RE GOING TO GET THAT CHANCE! GRIZZLY IS COMING THIS WAY!
MOVE WHILE YOU CAN, JIMMY! I'LL DISTRACT HIM WHILE YOU RUN!
SKOWW

ENOUGH, DOOMSDAY! IF YOU WANT TO GET YOUR HANDS ON MY FRIENDS--
--YOU'RE GOING TO HAVE TO KILL ME FIRST!
BARRH!
WHAT IS HE? WHAT DOES HE WANT?

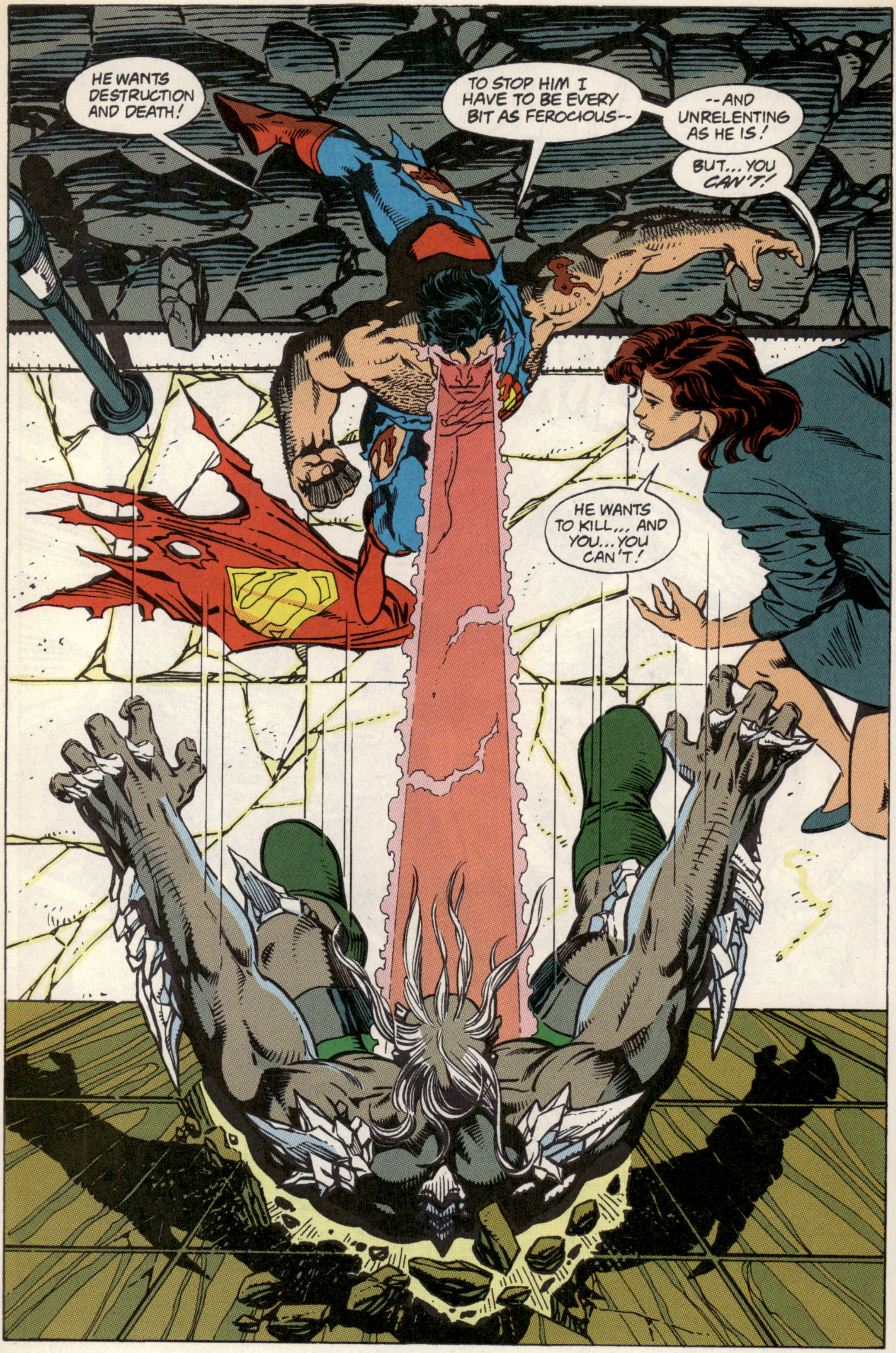
HE WANTS DESTRUCTION AND DEATH!
TO STOP HIM I HAVE TO BE EVERY BIT AS FEROCIOUS--
--AND UNRELENTING AS HE IS!
BUT... YOU CAN'T!
HE WANTS TO KILL... AND YOU... YOU CAN'T!

HYAA!
ARRR!
BONY PROTRUSIONS... SO SHARP... HE CUT ME!

HA HAAA!
I'M HURT BAD... CAN BARELY STAY CONSCIOUS!
UNNNGH!
MUST... TAKE HIM DOWN...
...NOW!

IF THOSE BONES ARE JUST EXTENSIONS OF HIS SKELETON--!
YES! I'VE FINALLY MANAGED TO HURT HIM!
SNAKKT
AAARRGH!
LUTHOR DONAT
BUILDING:

SKRASH
SKRASH
DAILY PLANET
SKRASH
SKRASH
BDAMMM
HUHN!
THEY HIT EACH OTHER SO HARD THE WINDOWS ARE SHATTERING!
EXHAUSTED... BUT I HAVE TO KEEP FIGHTING... UNTIL I DROP--
--OR HE DOES!
WATCH OUT FOR FALLING GLASS!

THIS IS IT! LOOKS LIKE WE'RE BOTH BETTING EVERYTHING WE'VE GOT ON THIS ONE!
FOR LOIS... AND JIMMY... FOR THIS ENTIRE CITY... I'VE GOT TO PUT THIS GUY AWAY WHILE I STILL CAN!
RRRRRRR

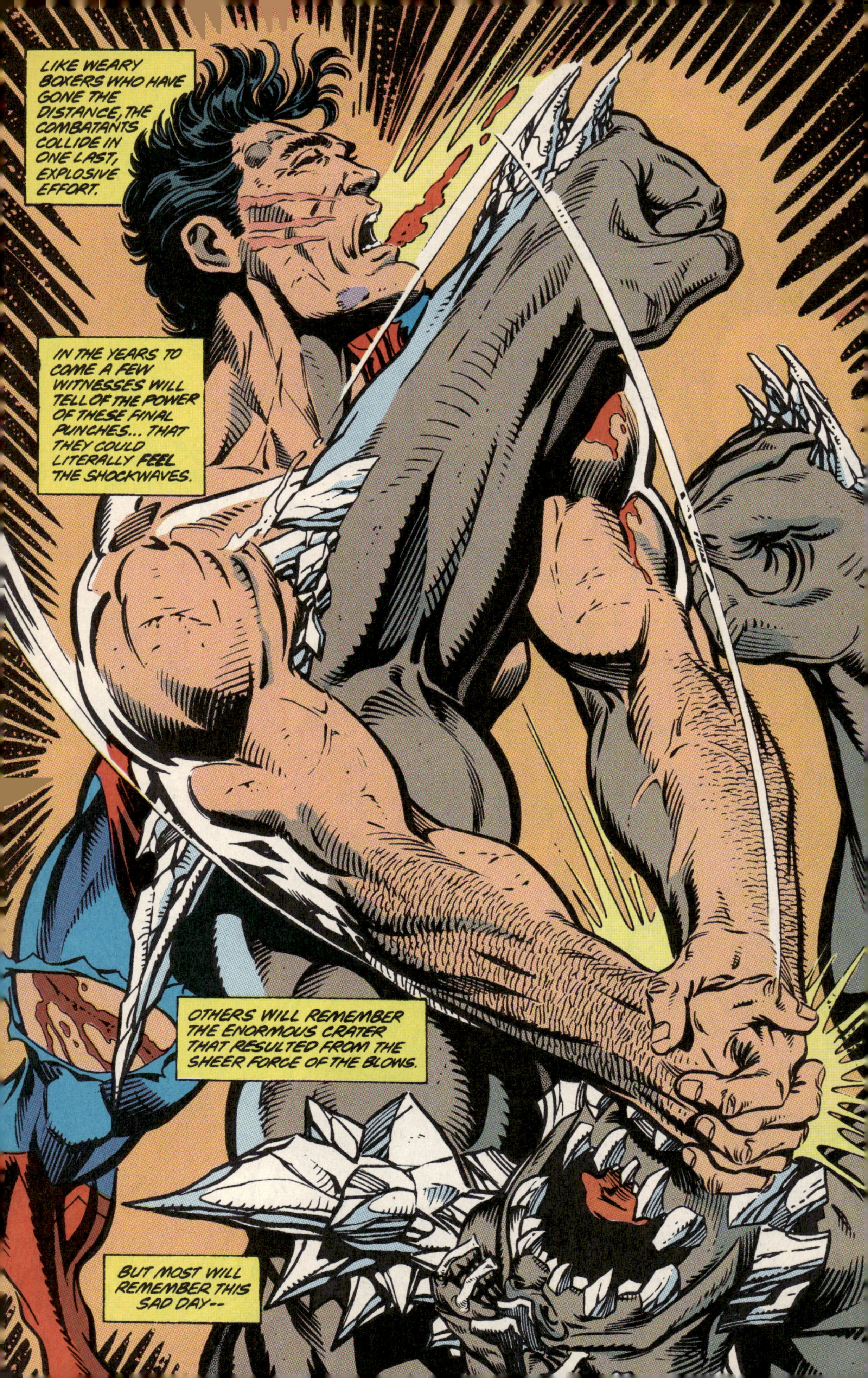
LIKE WEARY BOXERS WHO HAVE GONE THE DISTANCE, THE COMBATANTS COLLIDE IN ONE LAST, EXPLOSIVE EFFORT.
IN THE YEARS TO COME A FEW WITNESSES WILL TELL OF THE POWER OF THESE FINAL PUNCHES... THAT THEY COULD LITERALLY FEEL THE SHOCKWAVES.
OTHERS WILL REMEMBER THE ENORMOUS CRATER THAT RESULTED FROM THE SHEER FORCE OF THE BLOWS.
BUT MOST WILL REMEMBER THIS SAD DAY--

--AS THE DAY THE PROUDEST, MOST NOBLE MAN THEY EVER KNEW--
--FINALLY FELL.
NO.
FOR THOSE WHO LOVED HIM--
--ONE WHO WOULD CALL HIM HUSBAND--
--ONE WHO WOULD BE HIS PAL--
LEXCAM

--OR THOSE WHO WOULD CALL HIM SON--
--THIS IS THE DARKEST DAY THEY COULD EVER IMAGINE.
THEY RAISED HIM TO BE A HERO... TO KNOW THE VALUE OF SACRIFICE.
TO KNOW THE VALUE OF LIFE.

AND FOR THOSE WHO SERVED WITH SUPERMAN IN THE PROTECTION OF ALL LIFE--
--COMES THE SHOCK OF FAILURE.
THE WEIGHT OF BEING TOO LATE TO HELP.
WE... WE DIDN'T MAKE IT IN TIME!
HE MUST SURVIVE. IT CANNOT END LIKE THIS!
BUT IT WILL.
FOR A CITY TO LIVE, A MAN HAD GIVEN HIS ALL AND MORE.

DAILY PLANET

PLEASE HANG ON! THE PARAMEDICS WILL BE HERE ANY SECOND!
PLEASE!
DOOMSDAY... IS HE... IS HE...
YOU STOPPED HIM! YOU SAVED US ALL!
NOW RELAX UNTIL--
BUT IT'S TOO LATE.
FOR THIS IS THE DAY--

-- THAT A
SUPERMAN
DIED.